Great Inventors and Inventions

Bruce LaFontaine

DOVER PUBLICATIONS, INC.
Mineola, New York

Copyright

Copyright © 1997 by Dover Publications, Inc.
All rights reserved under Pan American and International Copyright Conventions.

Published in Canada by General Publishing Company, Ltd., 30 Lesmill Road, Don Mills, Toronto, Ontario.

Bibliographical Note

Great Inventors and Inventions is a new work, first published by Dover Publications, Inc., in 1997.

DOVER *Pictorial Archive* SERIES

This book belongs to the Dover Pictorial Archive Series. You may use the designs and illustrations for graphics and crafts applications, free and without special permission, provided that you include no more than four in the same publication or project. (For permission for additional use, please write to: Permissions Department, Dover Publications, Inc., 31 East 2nd Street, Mineola, N.Y. 11501.)

However, republication or reproduction of any illustration by any other graphic service, whether it be in a book or in any other design resource, is strictly prohibited.

International Standard Book Number: 0-486-29784-5

Manufactured in the United States of America
Dover Publications, Inc., 31 East 2nd Street, Mineola, N.Y. 11501

NOTE

No sharp distinction can be drawn between invention and discovery. A potent mixture of potassium nitrate, carbon, and sulfur, was almost certainly a discovery, yet we consider the resulting gunpowder an invention. The discovery that liquids expand and contract with variations in temperature led to the invention of the thermometer. Invention is a matter of application and scientific heritage; broadly defined, invention is an act of the imagination. Conversely, inventions can lead to discovery. From observation of the steam engine came theories that led to the development of thermodynamics, and to more efficient apparatuses.

Technology is about heritage. If not for a lack of technological heritage Leonardo da Vinci would most likely have invented a practical flying machine. For lack of this heritage, Thomas Edison might never have invented a thing. The difficulty of crediting any inventor with having been the first in his or her field is made clear when we look at the history of the steamboat. There were about 30 more or less successful steamers built before Robert Fulton went up the Hudson in his "Clermont" (1807). There was, for example, John Fitch who already had a steamboat run on the Delaware river in 1787; and William Symington, an Englishman whose "Charlotte Dundas" was a notable success (1802). Likewise, the inventions of the motion picture, television, the filament electric lamp, the sewing machine, and the elevator also shatter the pretensions of "first" inventors as a class. It follows that there are few inventions that owe nothing to earlier technology. Most of those that do can be attributed to primitive man: the wheel, the bow and arrow, the boat, the lever, etc. Modern inventions as a whole are combinations of well known standards that produce new results, or serve new purposes.

Described and depicted in this book are forty-five inventions and their inventors. Beginning with Gutenberg's movable type, noted illustrator Bruce LaFontaine—who specializes in the history of science and technology—presents historical details of important inventions, along with biographies of the inventors. Keeping the statement above in mind, and in the interest of revealing the process of scientific heritage, there is more than one inventor listed for some inventions.

JOHANNES GUTENBERG

MOVEABLE TYPE (1438) JOHANNES GUTENBERG

Before the invention of moveable metal type, books had to be hand-copied by scribes, a slow and laborious method, making books rare and very expensive. Johannes Gutenberg (1400–1468) is credited with inventing a much more efficient and economical way to produce books. In 1438 he developed a metal alloy that could be melted at low temperature, was easily molded into individual letters, and did not deform under the pressure of a press. This allowed him to produce individual metal type "slugs" with single letters upraised on the surface of the slug: using a composer's stick, he could arrange these type slugs into words, then into sentences, and then set them into a "form" which made up one complete page. While the form was held securely in place, a roller applied ink to the upraised letters. Then a sheet of paper was placed on top of the form and a modified wine press was used to ink the words onto the paper. New pages were created by rearranging individual type slugs into different words, sentences, and forms. From a single form, many copies of a page could be printed quickly and easily.

Gutenberg's print shop created the first complete book, the Bible, in 1455. When Gutenberg's printing system became known throughout Europe, books became less expensive, and thus more available to the common person. With this dramatic increase in the number of books, the desire for knowledge and learning developed rapidly. From this foundation, the "Renaissance" or "rebirth" emerged from the darkness and superstition of the medieval era, and with it came the beginnings of modern civilization.

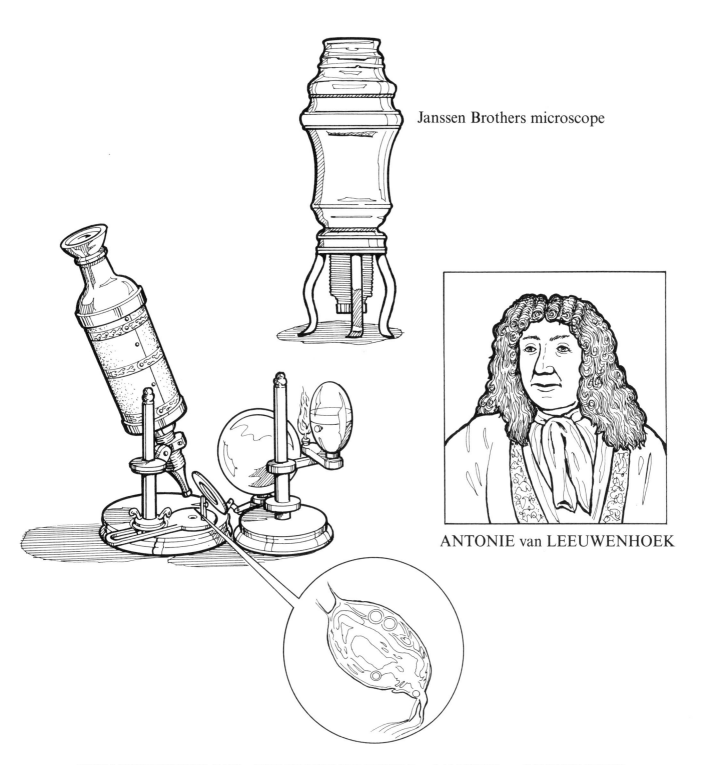

Janssen Brothers microscope

ANTONIE van LEEUWENHOEK

THE MICROSCOPE (1590)   THE JANSSEN BROTHERS and ANTONIE van LEEUWENHOEK

An important invention in the development of science and medicine was the microscope. It was based upon the principle that light could be "refracted," or bent, by a glass lens. It was soon discovered that tiny objects could be magnified in size when viewed through a glass lens that had been ground and polished in a specific manner. Although this principle was known to the Chinese as early as 1000 A.D., it was not until the 13th and 14th centuries in Europe that it was put to practical use in the form of eyeglasses.

In Europe the first microscope was invented by brothers Zacharias and Hans Janssen, two Dutch eyeglass-makers, around 1590. They built a "compound" microscope, so called because of its two lenses, one used to magnify the object, and the other to further enlarge the magnified image.

The most significant development and use of the microscope during this period, however, belongs to another Dutch optician, Antonie van Leeuwenhoek (1632–1723). Born in Delft, Holland, he became skilled at grinding very sharp and accurate magnifying lenses. Some of his single-lens microscopes were able to magnify up to three hundred times actual size, and around 1660 he began serious study using these instruments. He was the first to discover bacteria and other microscopic organisms, calling these tiny creatures "animalcules." Prior to this discovery, very small creatures such as fleas and maggots were thought to "spontaneously generate" from a single source, such as rotting meat in the case of maggots. He identified them as separate and distinct types of animals that had a reproduction cycle and did not spring into being magically. He was the first to discover the organisms that are present in saliva, blood and animal tissue. His exploration into the microscopic world was an invaluable aid to further study and discovery in science and medicine.

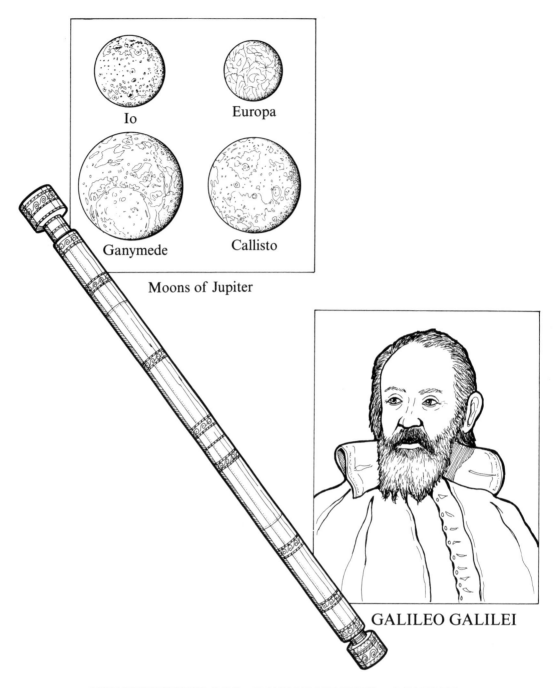

Moons of Jupiter

GALILEO GALILEI

THE TELESCOPE (1608)   HANS LIPPERSHEY and GALILEO

The foundation for the modern science of astronomy can be traced to the invention of the telescope. Using the same principle as the microscope, that a glass lens can "refract" light, Dutch eyeglass-maker Hans Lippershey (1570–1619) constructed the first telescope in 1608. It consisted of two lenses held in line within a tube. One lens was fixed and the other movable, to provide focus. He called the device a *kijker,* or viewer.

As in the case of the microscope, the use and development of the telescope was advanced by a scientist other than its inventor. Galileo Galilei (1564–1642), one of the preeminent scientists of the Renaissance and one of the greatest minds in history, made critical and fundamental discoveries with his telescopic observations.

Galileo was already a well-known scholar and scientist when he constructed his first telescope, based on Lippershey's design, in 1609. He soon began his study of the night sky and recorded his observations with careful drawings. He was able to see mountains and craters on the moon, previously unseen. He discovered four large moons circling the planet Jupiter. These and other discoveries convinced Galileo that the accepted theory of the basic nature of the solar system was incorrect. At that time it was still believed that the Earth was the center of our solar system (and indeed, the entire universe), and that the sun, other planets, and all celestial bodies revolved around our planet. Galileo's observations gave credible physical proof that another theory was true, one put forth by Polish astronomer Nicolaus Copernicus in 1543. This theory correctly deduced that the sun was the center of our solar system and that all the planets and their moons revolved around it. Because this theory was opposed by the powerful Roman Catholic Church, Galileo was accused of heresy (religious treason, a very serious crime at the time). He was ordered by the Church tribunal to "recant" his belief—that is, to admit he was wrong—under penalty of death. He gave in to this very real threat and was allowed to live. Nevertheless, other scholars and scientists were convinced that he was right, and the sun-centered theory of the solar system gained wide acceptance.

Galileo's observations and discoveries, as well as the controversy surrounding his religious trial, contributed greatly to the spread of new scientific knowledge throughout the world of the Renaissance.

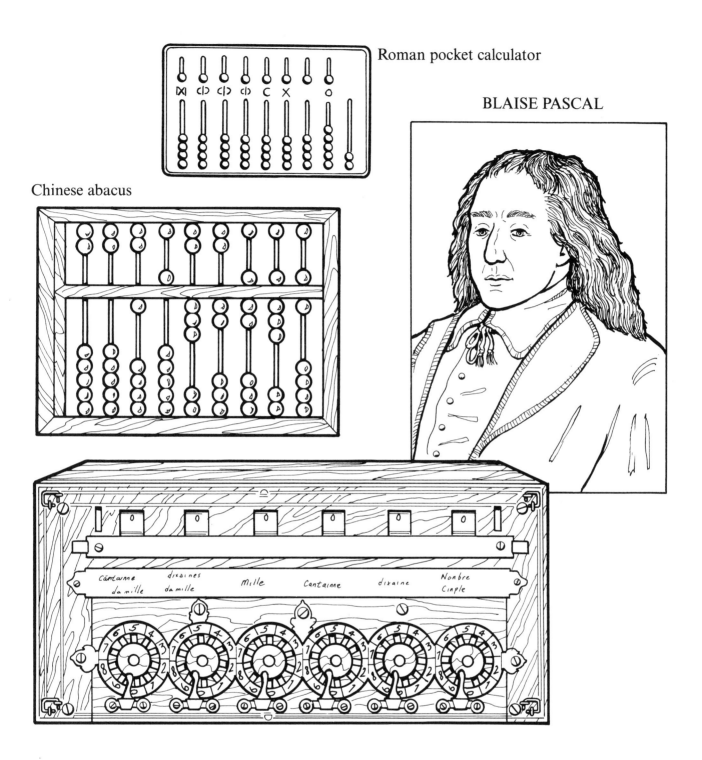

THE MECHANICAL CALCULATOR (1643) BLAISE PASCAL and CHARLES BABBAGE

A system to replace the time-consuming and tedious task of calculation by hand can be found as far back as 4000 years ago. The Chinese invented a device around 2000 B.C. called the abacus. It was a mechanism with wooden beads mounted in a frame. The beads represented hundredths, tenths, and other numerical units, and could be moved to various positions. Not quite as ancient, but still in the distant past, was a tool devised by the Romans called a "counting board," consisting of brass beads strung on rods. The upper beads represented five times the value of the lower beads. Unfortunately, these devices were efficient only in the hands of trained and skilled operators, making their practical use very limited.

The ancestor of the modern computer can be traced back to the year 1643 with a mechanical calculating machine built by French mathematician and philosopher Blaise Pascal (1623–1662). Although they were extremely difficult for the instrument makers of that era to build, around seventy of Pascal's calculating machines were eventually made.

Another, much-later mathematician and inventor who attempted to build a calculating machine was Charles Babbage (1792–1871). In 1843 he began work on what he called an "analytical engine." The device was extremely complicated, with thousands of intricate parts. Even though the tool and instrument-makers of his era were more skilled than in Pascal's day, they were never able to make the parts to the exacting tolerances needed for the machine to work properly. Babbage spent thirty-seven years trying unsuccessfully to complete his "analytical engine." It was not until the advent of modern electronics in the 1940s that the direct predecessor of modern computers, ENIAC, was built.

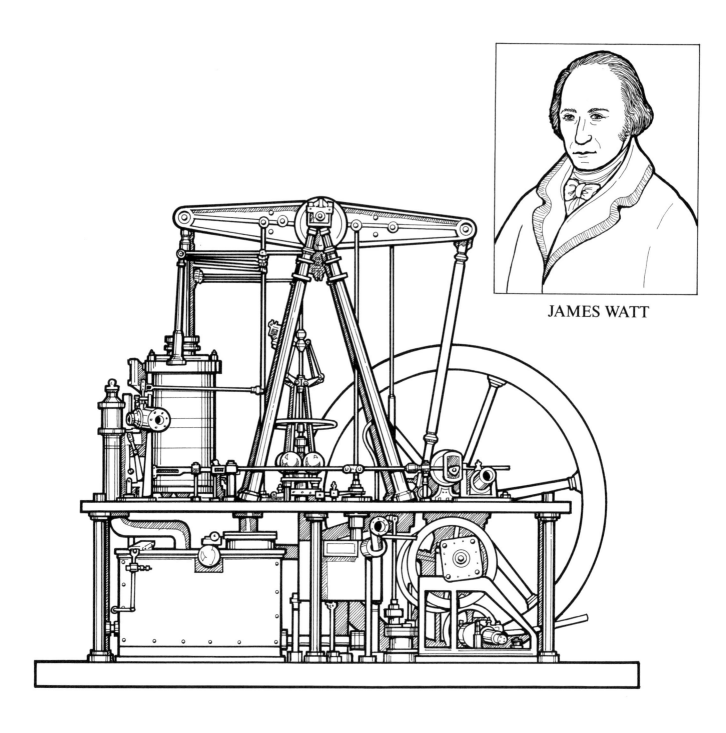

THE STEAM ENGINE (1712) THOMAS NEWCOMEN and JAMES WATT

The steam engine was based on the principle that water heated in a closed chamber will turn to steam and begin to expand in volume. This expansion causes pressure which can be used to move a piston within a cylinder. This movement can be used in either an up-and-down or rotary (circular) motion to power and operate a wide range of machines.

Englishman Thomas Newcomen (1663–1729) was the first to design and build a working steam engine around 1712. Created to operate pumps to remove water from Welsh coal mines, these first steam engines were crude and inefficient, but they were soon adapted for use throughout Europe. Again, as in the case of the microscope and telescope, the originator of the steam engine was eclipsed in historical significance by a later inventor. When he refined and improved upon Newcomen's design, Scottish instrument-maker James Watt (1736–1819) turned the steam engine into a reliable, fuel-efficient power plant, and for doing so is credited as being the "father of the steam engine." Watt's primary contribution to the technology was the invention of the condenser in 1769. This was a separate apparatus which allowed steam to quickly condense back into water, to be yet again reheated into steam. Other improvements Watt made to the steam engine were: a piston shaft with rotary motion instead of up-and-down movement, allowing for a wider range of applications; a double-action piston enabling steam to enter alternately from the top and bottom of the cylinder, thus increasing the engine's working capacity; and a speed governor to regulate the engine speed by controlling the flow of steam.

By the end of the 18th century over five hundred steam engines designed by Watt were in use. They were a key part of the transformation of our society from an agrarian, or farm-based, culture to an industrial civilization, and provided the power for the inventions of the 19th century yet to come.

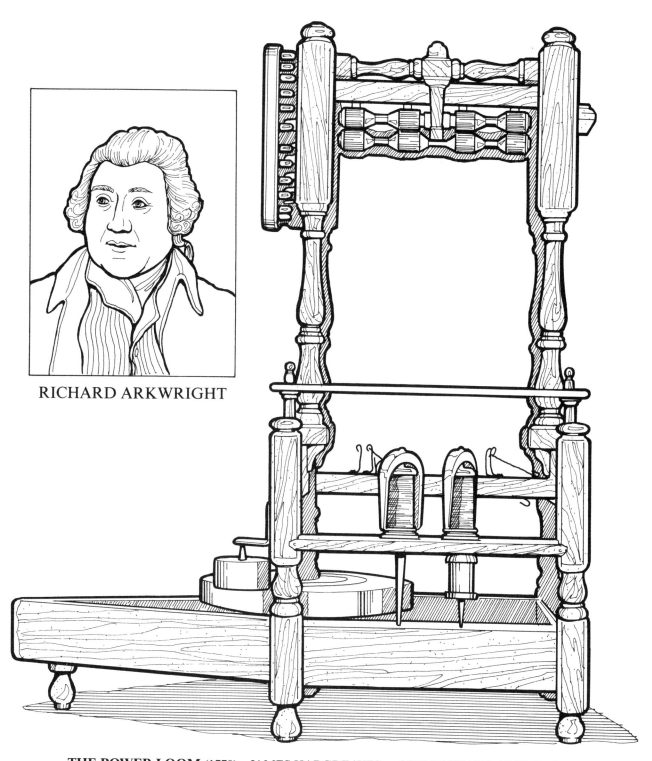

RICHARD ARKWRIGHT

THE POWER LOOM (1770)   JAMES HARGREAVES and SIR RICHARD ARKWRIGHT

One of the important inventions that contributed to the industrial development of Great Britain and Europe was the "Spinning Jenny," a mechanized spinning machine invented by James Hargreaves (1722–1778). This device, invented in 1765, spun raw wool, linen, and cotton fiber into thread for use in weaving cloth. With its many spindles it was able to do the work of many individuals operating single spinning wheels by hand. These workers were part of a cottage industry (individuals who worked out of their homes, or cottages) in rural districts of England. When word of Hargreaves new machine became widely known these workers felt threatened with the loss of their jobs and livelihood. Riots broke out and Hargreaves' home was burned to the ground and his equipment destroyed. Hargreaves was not deterred by this event; he simply moved to another district and rebuilt his machines. His invention became widely used and made a permanent change in the spinning industry.

In 1770, a few years after the appearance of Hargreaves' spinning machine, Sir Richard Arkwright (1732–1792) invented the water frame power loom, so called because it was operated by the power of a water wheel. His water frame loom produced a stronger thread than hand-powered looms and was so efficient that factories with many power looms were soon established throughout England. In 1785 Arkwright was the first to utilize one of James Watt's steam engines to drive a cotton mill, another step in increasing productivity. His power loom and factory system created the foundation of a major industry in Great Britain, cloth-making, which by the 1840s accounted for over forty per cent of British exports.

THE SUBMARINE (1776) DAVID BUSHNELL

To explore the mysteries of the deep has been an age-old desire of mankind. The quest for a submersible boat, or "submarine," was also motivated by a desire for a stealth weapon for use in naval warfare.

In 1776 an American inventor, David Bushnell (1742–1824), designed and built an egg-shaped submarine called *Bushnell's Turtle*. It was intended for use against British warships blockading New York harbor during the American Revolutionary War. The *Turtle*, constructed from wooden planks with iron and brass reinforcements, carried a single crewman who hand cranked propellers for propulsion. The tiny craft had an air supply that allowed it to remain submerged for thirty minutes. A wooden mine containing 150 pounds of gunpowder was its only armament. The mine was to be attached to the hull of a British ship with a screw device, and then detonated from a safe distance. On the night of September 6, 1776, Sergeant Ezra Lee of the Continental Army piloted the *Turtle* to the hull of the British ship *Eagle*. Although the submarine worked well, Sergeant Lee was unable to attach the bomb because of the copper plating covering the hull of the *Eagle*.

Another American inventor, Robert Fulton (1765–1815), designed and built a submarine called the *Nautilus*, in 1800. He was unable to secure government support, however, and the project ended after just a few favorable test dives.

A more successful attempt at submarine warfare was made during the American Civil War. In 1864 the Confederate forces launched an all-metal submersible named the *Hunley* against the Union ship *Housatonic,* anchored in Charleston harbor. It successfully guided a torpedo to the *Housatonic*, sinking the ship. Unfortunately, the waves caused by the explosion swamped and flooded the *Hunley* through an open hatch, and the craft sank with all hands lost.

In 1875, American naval architect John Holland (1840–1914) began the design and construction of a series of submarine prototypes culminating in the successful launching of the *SS-1*, accepted by the U.S. Navy in 1900. In its basic features, the *SS-1* was the model upon which modern submarines were developed.

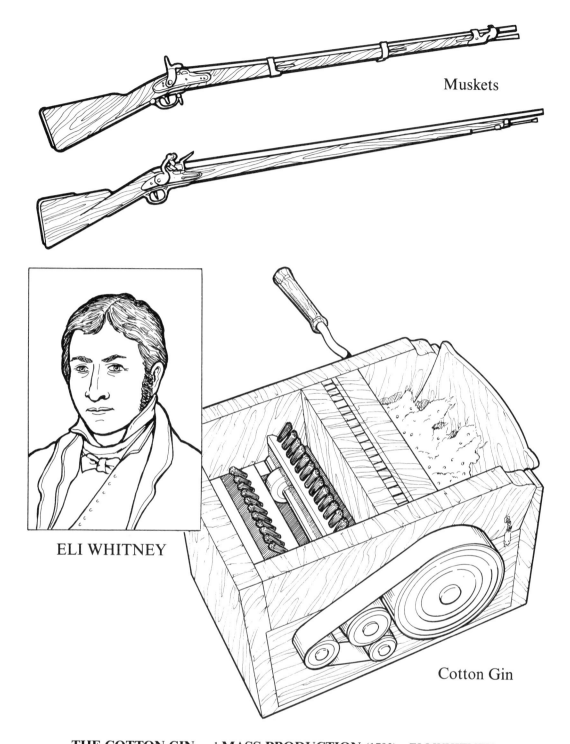

THE COTTON GIN and MASS PRODUCTION (1793)   ELI WHITNEY

Eli Whitney (1765–1825) was an important American inventor whose ideas were indispensable to the early economic development of the United States. His invention of the cotton gin in 1793, enabled the cotton-growers of the American South to revolutionize agriculture. His device cleaned the seeds out of fifty pounds of raw cotton per day—a necessary procedure before it could be further processed into thread and cloth—far faster than manual processing, which could produce only one pound of cleaned cotton every three hours.

Whitney was also instrumental in demonstrating the advantages of mass production with the use of standardized, interchangeable parts. When he received a contract from the U.S. government in 1797 to build 10,000 muskets within two years in an era when muskets were individually hand-crafted by skilled gunsmiths producing a maximum of 250 guns per year, he needed a new method to accomplish the task. Whitney established a factory in Connecticut based on the idea that every individual part of a single musket could be interchanged with a like part from a similar musket. By doing this he could hire many more unskilled workers to assemble the interchangeable parts, enabling his factory to drastically increase production and meet the government's requirement for 10,000 guns.

With the success of this system of manufacture, many other industries adopted mass production for their products, including clock makers, and tool and instrument makers. Another American industrial pioneer, Henry Ford, would later use mass production in an even more effective and significant way to lay the foundations of the American automobile industry.

ROBERT FULTON

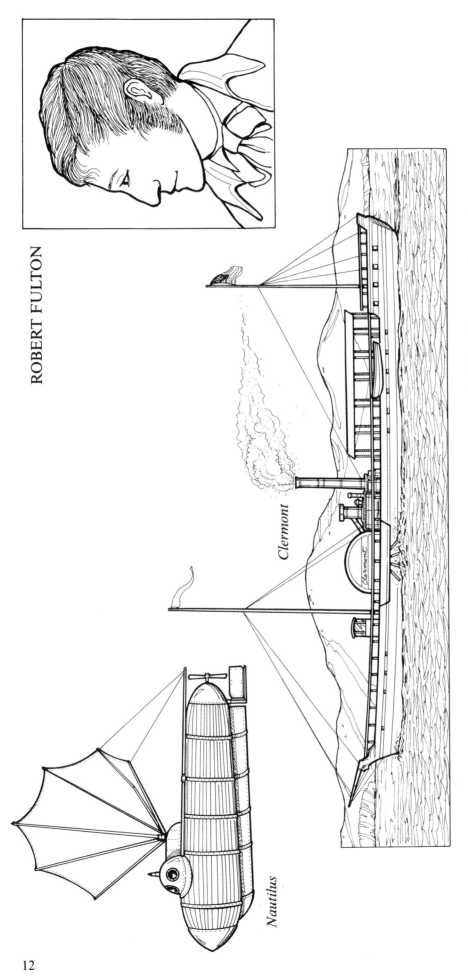

Clermont

Nautilus

THE STEAMBOAT (1783)   JOHN FITCH, WILLIAM SYMINGTON, and ROBERT FULTON

A significant development in the use of the steam engine was its application aboard ships. An early attempt at a steam driven ship was made in 1783 by American John Fitch (1743–1798). He devised a system that enabled a steam engine to propel a ship through the water using oars. Through a complex mechanical collection of levers, pulleys, and ropes, the oars were able to move the small craft at a very slow rate of speed. Because of the inefficiency of the mechanism, the project was not developed any further. A more successful attempt at a steamship was made in 1802 by British engineer William Symington (1763–1831). His vessel, the *Charlotte Dundas* used paddle wheels for propulsion and worked as a tugboat on the Scottish canal system for a number of years.

American inventor Robert Fulton (1765–1815) is generally credited with creating the first passenger and freight carrying steamboat to go into regular service. The *Clermont* (1807), his 150 foot long craft, was propelled by a paddle wheel on either side of the hull and successfully traveled the Hudson River between New York City and Albany in thirty-two hours. The success of the *Clermont* inspired shipbuilders worldwide and the paddle wheel and propeller driven steamer soon became a commonplace sight on the oceans and rivers of the world. A few years after the *Clermont*, in 1812, Fulton also built the first steam powered warship for the U.S. Navy. Called the *Demologos* it was used against the British fleet during the War of 1812.

Fulton was also a pioneer in the development of the submarine. In 1800 he designed and built a twenty-one foot long submersible boat for the French government. It was constructed of copper and wood and featured a collapsible sail for surface propulsion. A hand cranked propeller was used for power. Despite making several successful test dives, it was not developed any further by the French. Nonetheless, its innovative design contained many features found on modern submarines including diving planes, flooding valves and chambers, and a conning tower. The *Nautilus*, as it was christened by Fulton, was to sail again in fictional form in Jules Verne's 1870 novel, *20,000 Leagues Under the Sea*. The world's first nuclear powered submarine was also christened the *Nautilus* in 1954, by the U.S. Navy.

ALESSANDRO VOLTA

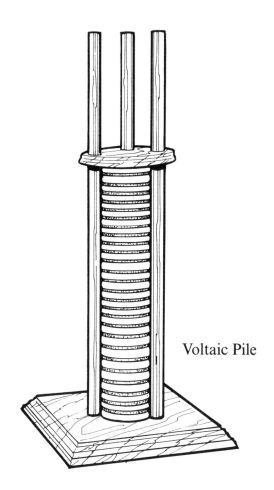

Voltaic Pile

W.H. Wollaston's battery

THE ELECTRIC BATTERY (1794)   ALESSANDRO VOLTA

To unlock the secrets of the mysterious force of electricity was a major goal for many scientists of the late eighteenth century. Three pioneers in this early research were Englishmen Humphrey Davy (1778–1829), Michael Faraday (1791–1867), and Italian physicist Alessandro Volta (1745–1827).

Humphrey Davy discovered the process of "electrolysis," whereby a chemical compound could be broken down into its constituent elements when an electrical charge was passed through it. The simplest example of this is water being reduced to its two basic elements, hydrogen and oxygen. Michael Faraday invented the electric motor, and the dynamo, a device which generated electricity. Alessandro Volta invented the electric battery, whose basic principle of operation remains the same today.

Volta began his work by creating a device that could store a static electrical charge called an "electrophorus" in 1775. This formed the basis for a fundamental piece of electrical equipment called a "condenser," a mechanism to store electricity within electrical circuits. It is still an indispensable component of modern electrical devices.

In 1794 Volta discovered that if two metals were brought together in a salt solution, an electric current would be generated. He constructed a series of alternating copper and zinc discs separated by felt pads soaked in salt water. His device was known as a "Voltaic pile," now called a battery. It was able to create and store electrical energy.

Other scientists of this period were soon experimenting with their own battery concepts and the device became an invaluable aid in conducting electrical experiments. A common type in use around this time was the "bucket battery" (shown above). Invented by Englishman W.H. Wollaston in 1807, it featured alternating metal plates of copper and zinc, suspended in a salt solution of ammonium chloride.

For his achievements in electrical research, Volta was honored by the Emperor Napoleon with the Legion of Honor and granted the title of Count. He will best be remembered, however, for the fact that the basic unit of electrical energy, the volt, was named in his honor.

JOSEPH-MARIE JACQUARD

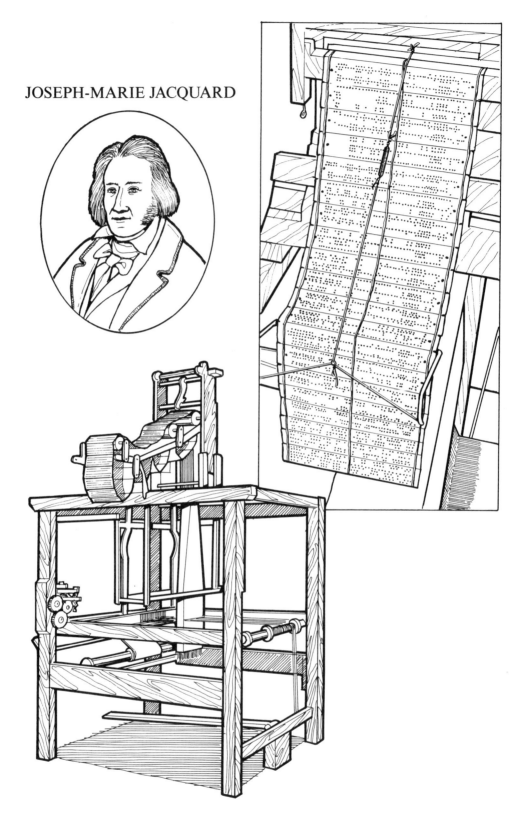

THE JACQUARD AUTOMATIC LOOM (1805) JOSEPH-MARIE JACQUARD

An important development in the growth of the textile industry was the automatic power loom, invented in 1805 by Joseph-Marie Jacquard. It was controlled by a series of paper cards with holes punched in them, and was able to weave complex patterns reliably. The threads of the loom were moved by sprung needles that lifted only those threads corresponding to the punched pattern on the cards. In this way, weavers were able to increase productivity and reduce human error in the difficult task of reproducing complicated textile designs. The loom itself was water-powered.

Much like the workers in the English spinning industry of the 18th century, the silk weavers of nineteenth century France saw the automatic loom as a threat to their livelihood. An angry mob of workers attacked Jacquard on the street and destroyed his equipment. The French government stepped in and placed his machine in the public domain, making it available to individual workers in the weaving industry. As recognition for his invention of the automatic loom, the French government awarded Jacquard a royalty from the cost of each loom sold, and he became a wealthy man.

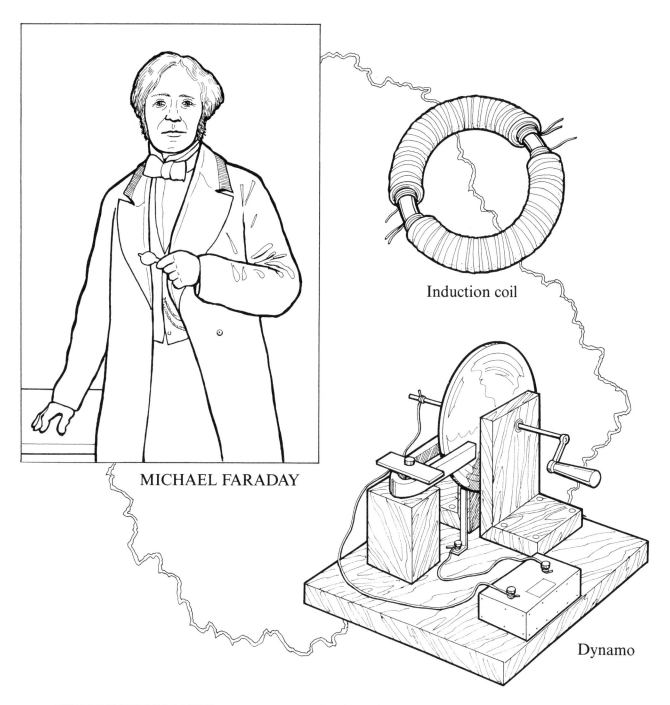

Induction coil

MICHAEL FARADAY

Dynamo

THE ELECTRIC MOTOR (1821) and THE ELECTRIC DYNAMO (1831) MICHAEL FARADAY

Michael Faraday (1791–1867) was a self-educated English scientist and inventor credited with inventing both the electric motor and the electric dynamo. He was born into a poor family of ten children, and apprenticed to a firm of booksellers at the age of fourteen. His employers encouraged him to read any of the books available in the store, and he developed an intense interest in the new science of electricity. In 1813, after a series of correspondence with Humphrey Davy, the foremost electrical expert of the day, he was hired by Davy as his assistant. Faraday then began his lifelong research and experimentation on the properties of electricity.

Faraday was inspired by a discovery made by Danish physicist Hans Oersted that the needle of a compass would be deflected from true North when a wire with electric current was passed over the compass. With this in mind, Faraday began the investigation of the relationship between electricity and magnetism. In 1821 he constructed a simple device consisting of a wire with electric current running through it which was placed between the poles of a horseshoe magnet. The interaction of electrical and magnetic forces caused the wire to rotate. Faraday had produced mechanical motion from an electric current, the basic principle of the electric motor.

His next invention was the induction coil. This apparatus "induced" an electric current by means of fluctuating fields of magnetic force, which were created by wrapping insulated wire around a ring of soft iron, which in turn was wrapped in silk. This simple mechanism opened the door to the mathematical understanding of the interaction between electricity and magnetism, now known as the Electromagnetic Theory–considered one of the fundamental laws of our physical universe. In 1831 Faraday built the first "dynamo," or electrical generator. He placed a copper disc with a crank handle between the two poles of a horseshoe magnet. By spinning the disc between the magnets he was able to produce continuous electric current. By his experimentation with these crude electrical devices, Faraday provided the foundation for hundreds of electrically powered inventions created later.

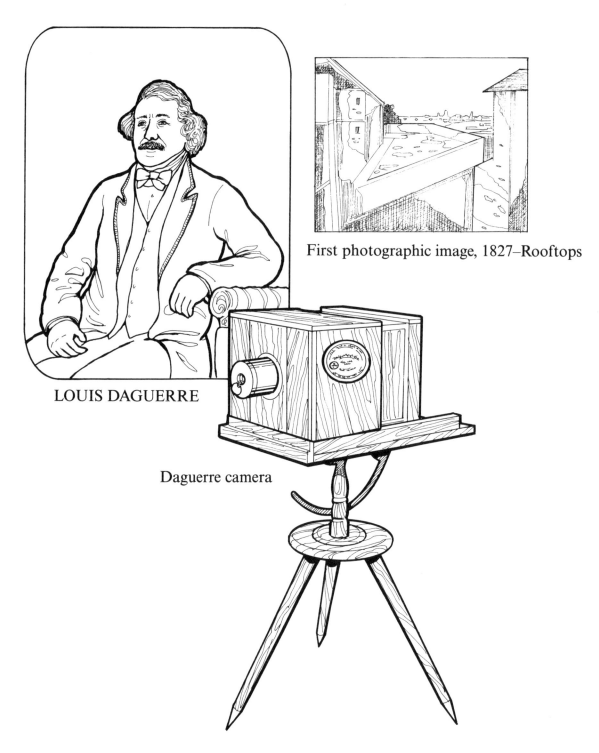

LOUIS DAGUERRE

First photographic image, 1827–Rooftops

Daguerre camera

PHOTOGRAPHY (1826)
JOSEPH-NICEPHORE NIEPCE, LOUIS-JACQUES-MANDE DAGUERRE, and GEORGE EASTMAN

Renaissance painters were so skilled at rendering the physical world in oil paint, that their work is often hard to distinguish from a photograph. Their technical ability to manipulate light—defined by the term chiaroscuro— has never been surpassed. But the actual capacity to create and preserve images from the real world was not attained until 1826.

A French inventor, Joseph-Nicephore Niepce (1765–1833), working with a simple camera of his own design, created the first photographic image. It showed a hazy but recognizable view from his rooftop. He had achieved this result after experimenting with coating a metal plate with different light sensitive chemicals, then exposing the plate by direct sunlight through his camera lens. He called this process heliography. Although it took seven hours of actual exposure time, he was finally able to attain a permanent image on the light sensitive plate.

His work was admired by another French inventor, Louis-Jacques-Mande Daguerre (1789–1851), who persuaded Niepce to become his partner in 1829. Daguerre was an energetic and innovative photographic experimenter and eventually improved upon Niepce's methods, inventing his own more practical system of capturing photographic images. In 1837 he discovered the right mix of chemicals to develop and fix an image on a metal plate with an exposure time of only twenty minutes. His photographs were clear and sharp and his process became widely used by other photographers. By 1850, English photographer William Henry Fox Talbot had advanced the science of photography by using better camera

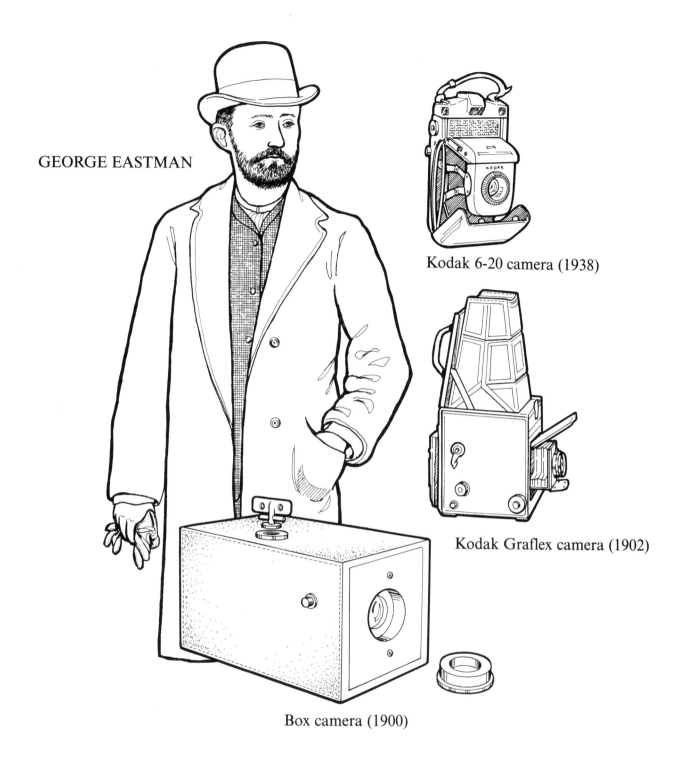

GEORGE EASTMAN

Kodak 6-20 camera (1938)

Kodak Graflex camera (1902)

Box camera (1900)

lenses and more sensitive chemicals, and by inventing the present day system of creating a negative image first and then producing a positive print from that.

American inventor George Eastman (1854–1932) took the next step forward in photography by making it available to the public when he introduced flexible celluloid roll film and the "Kodak" box camera in 1888. This much simpler system replaced the cumbersome metal and glass plates previously used. Eastman's improvement was simple, a reliable and inexpensive box camera that came already loaded with 100 exposures of his roll film. When all the exposures were made, the entire camera was returned to the Kodak factory in Rochester, New York, where it was developed, printed, and the camera reloaded. Eastman's concept of one step photography was marketed with the slogan, "You press the button, we do the rest," and set the standard for the new photographic industry. Eastman also preceded Henry Ford in the concept of mass production by setting up assembly lines to manufacture his cameras and film. He also worked with Thomas Edison to develop a celluloid based roll film for Edison's "kinetograph" moving picture camera. Eastman Kodak, as it eventually became known, further expanded the photographic market with the introduction of the "Brownie" camera for children in 1900.

George Eastman and his "Kodak" camera and film brought the popular hobby of photography to the world at large and in doing so created an international, multi-billion dollar industry. His enterprise also transformed the sleepy mill town of Rochester, New York, into a major technical and industrial city known worldwide as a center for optics and imaging science.

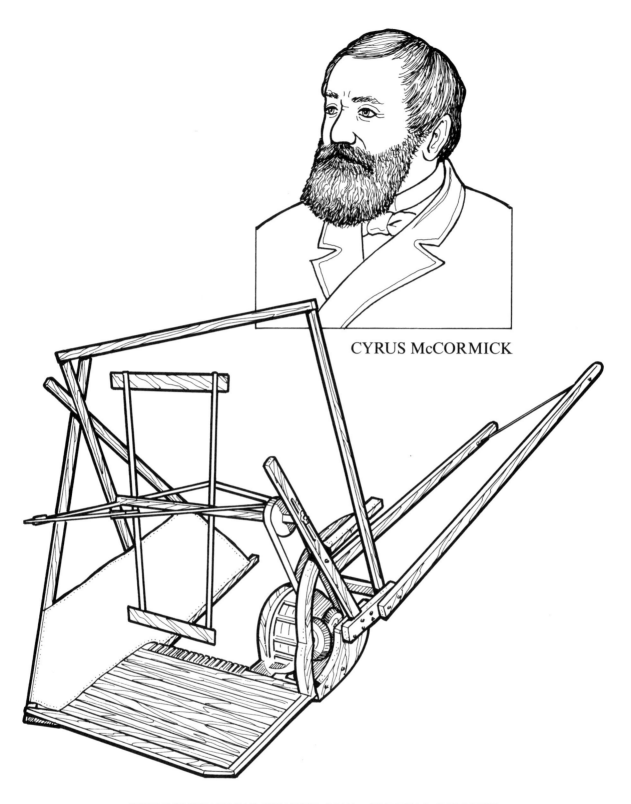

THE MECHANICAL REAPER (1831)   CYRUS McCORMICK

A significant contribution to the economic growth of the United States was the development of the vast agricultural potential of the American continent. An important element in this growth was the invention of the mechanical reaper by Cyrus McCormick (1809–1884). He built his first version of this machine in 1831. It was basically a wooden frame pulled by a team of horses. The device featured rotating blades that quickly and efficiently cut down fields of wheat at harvest time. Over the next twenty years he built successively better versions of his design until introducing his most advanced model in 1851 at the Great Exhibition of London. Farmers and agricultural experts were very impressed with the new machine and orders began to pour in. McCormick's factory was soon producing his mechanical reaper at a rate of 4000 machines per year.

By vastly improving the speed and productivity of harvesting crops, McCormick's machine was crucial in large-scale cultivation of immense tracts of open grassland in the American Midwest, which gained a reputation as the "bread basket of the world."

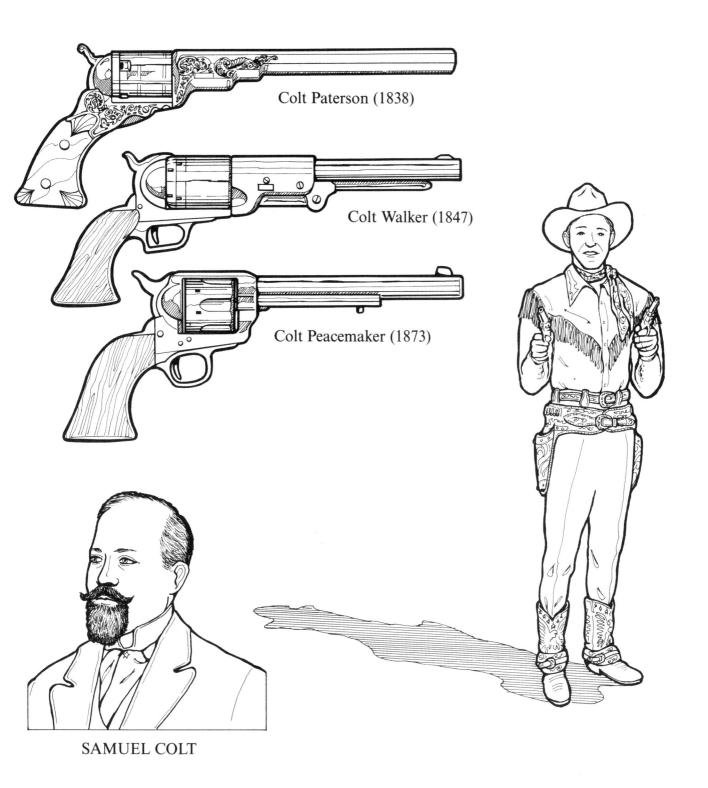

SAMUEL COLT

THE REVOLVER PISTOL (1838) SAMUEL COLT

The invention of the revolver pistol, first patented in 1838 by American inventor Samuel Colt (1814–1862), allowed for the rapid fire of multiple bullets or "rounds" of ammunition. Its basic mechanism was a rotating cylinder containing the ammunition, usually six bullets. Each time the trigger was pulled to fire a round, the cylinder moved to place the next bullet into firing position.

The revolver model shown at the top of the illustration is the Colt Paterson, built in 1838, which fired a .36 caliber bullet. Caliber is a designation indicating the size of the bullet. The larger the caliber, the more powerful the weapon. You will notice that the Paterson has no apparent trigger. In this particular weapon, the trigger was extended when the hammer at the back of the cylinder was pulled back, or cocked, in the ready to fire position. After pressing the extended trigger and firing the bullet, the trigger slid back into the handle.

The next model depicted is the Colt "Colonel Walker" Army revolver of 1847, a .44 caliber pistol. The final weapon shown is the famous Colt .45 caliber "Peacemaker" of 1873, often referred to as "the gun that tamed the West." It was used by legendary lawmen and cowboy heroes of the American West like Wyatt Earp, Wild Bill Hickok, and Buffalo Bill Cody, and it was equally made famous by their outlaw adversaries including Jesse James, Johnnie Ringo, and Billy the Kid, all part of a rich history bound together by the lore of the "six-gun."

SAMUEL F.B. MORSE

THE TELEGRAPH (1837)   SAMUEL F. B. MORSE

Another invention that played a key role in the opening of the American West was the telegraph. It allowed for instantaneous communications over the 3000 mile wide expanse of the American continent. This was crucial to enabling the settled and developed Eastern seaboard to maintain contact with the frontier wilderness of the far west.

The first practical telegraphic system was created in 1837 by an American artist and inventor, Samuel F. B. Morse (1791–1872). A well-known portrait painter whose clients included such famous men as Eli Whitney and the Marquis de Lafayette, he was also a gifted amateur scientist with a serious interest in electromagnetism. Although two British inventors, Charles Wheatstone and William Cooke, also developed their own telegraphic system in 1837, Morse's system proved to be the more efficient.

In its basic form, the telegraph consisted of a "transmitter" that sent electrical impulses over a metal wire to a recipient, using a "receiver." These impulses were decoded by using a system devised by Morse in 1838, now called "Morse Code." This code uses long "dashes" of electrical current, combined with short "dots" of current. Specific combinations of these dashes and dots represent letters of the alphabet and numbers. Words and sentences could then be created by properly intermixing these signals.

In 1844 the first permanent telegraphic line was built between Baltimore and Washington, D.C., a distance of thirty-seven miles. It consisted of wooden poles strung with iron wire that was insulated with glass door knobs. The first message was a famous quote which described the wonder and amazement at this new invention, "What hath God wrought?" Shown above is a depiction of Morse's prototype telegraphic transmitter built in 1837 using parts from a picture frame. Also shown is a more advanced commercial telegraphic unit built by the Siemens Company in 1856.

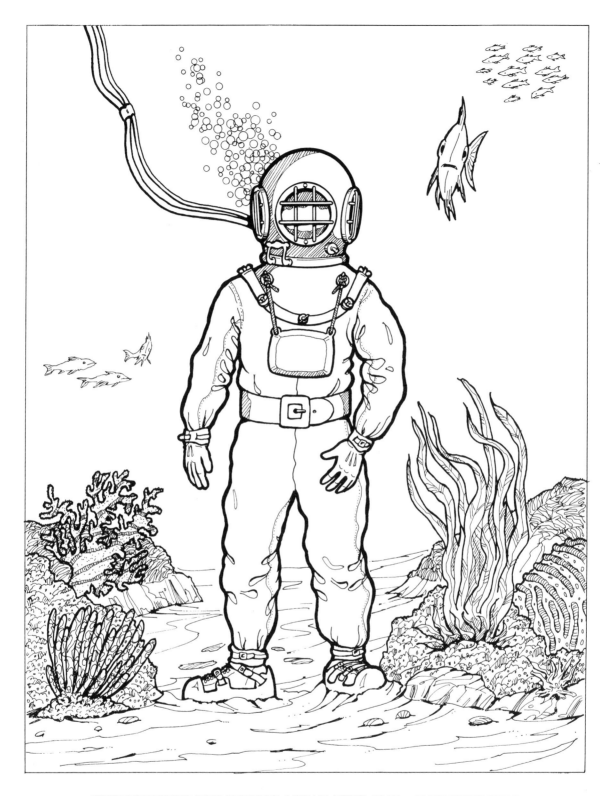

THE UNDERWATER DIVING APPARATUS (1837)   AUGUSTUS SIEBE

Attempts at devising a diving apparatus and remaining underwater for extended periods can be traced as far back as 900 B.C. Assyrian wall-relief sculptures have been found that depict men walking beneath the sea wearing goatskin bags attached to breathing tubes. It was not until 1819, however, that a successful diving system was invented by English mechanic Augustus Siebe (1788–1872). His "open-dress diving suit" had a copper helmet attached to a jacket reaching to the waist. Fresh air was pumped to the helmet from a surface ship and exhausted through the bottom of the jacket.

In 1837 Siebe improved upon his design with the introduction of his "closed-system diving suit," completely encasing the diver in canvas and rubber. This suit and helmet, shown in the illustration above, became the model upon which all successive diving gear was based well into the 1950's. A copper or brass helmet was mounted on a breastplate, weights and lead-soled boots allowed the diver to remain in an upright position and maintain stability when walking on the sea floor. In this design air was exhausted through a valve in the helmet. In 1943, another French inventor, Jacques-Yves Cousteau, created a new diving system called "SCUBA," short for Self-Contained Underwater Breathing Apparatus, which supplanted the "hard hat" diving system almost completely.

ELIAS HOWE

ISAAC SINGER

THE SEWING MACHINE (1845) ELIAS HOWE and ISAAC SINGER

During the nineteenth century two American inventors competed fiercely for the right to mass market the first sewing machines for home and industrial use. They were Elias Howe (1819–1867), who patented his design in 1845, and Isaac Singer (1811–1875), who built his machine in 1851.

Although Howe was the original inventor, he was unable to attract investors to finance mass production of his machines. He eventually moved to Great Britain where he sold his patent rights. In the meantime, Singer had improved upon Howe's original design with a machine of his own. With his partner Edward Clark, he was able to set up a factory and produce his machines in quantity. The Singer machines were very efficient and successful, and by 1860 the Singer & Clark company had become the world's largest manufacturer of sewing machines.

During this period Elias Howe had moved back to the U.S. with a lawsuit against Singer for patent infringement. In 1854 after several years in the courts, Howe won the lawsuit and was awarded royalties on all sewing machines sold in the United States. The invention transformed the world's clothing industry from individual seamstresses and tailors to mass production.

LENOIR BROTHERS

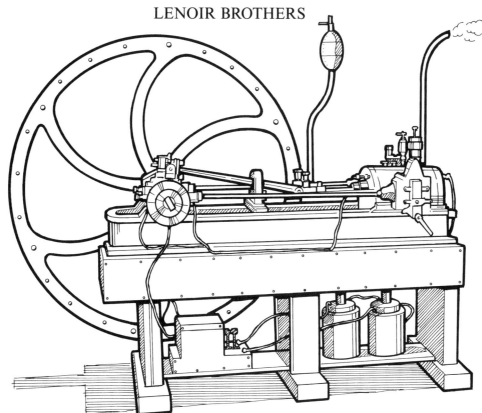

THE INTERNAL COMBUSTION ENGINE (1860) ETIENNE LENOIR

Successor to the steam engine as the world's most widely used power source, was the internal combustion engine. It was invented in 1860 by Belgian-born engineer Etienne Lenoir (1822–1900).

Lenoir's first version was basically a modified steam engine. It worked on the principle of the two-stroke cycle of combustion. Its fuel was a flammable gaseous substance derived from coal, appropriately called "coal gas." In Lenoir's design the coal gas and air were drawn into a cylinder by the momentum of a large flywheel. When the piston within the cylinder was part way down, an electrical spark ignited the fuel-air mixture causing an explosion. The expanding gases caused by the explosion forced the piston to complete its up and down stroke, while the exhausted gases were ejected through a valve in the cylinder, and the motion of the flywheel caused the cycle to begin again. Lenoir continued to improve and develop his engine and it was eventually modified to run on liquid fuel such as turpentine, kerosene, and finally, gasoline.

The advantages of the internal combustion engine over the steam engine are its greater efficiency and a much higher power-to-weight ratio. This meant that a much lighter engine could be used to produce the same or greater power. This fact was crucial in the selection and use of the internal combustion engine by the numerous inventors of the 1880s who were tinkering with a new machine called the "horseless carriage," later to be known as the automobile.

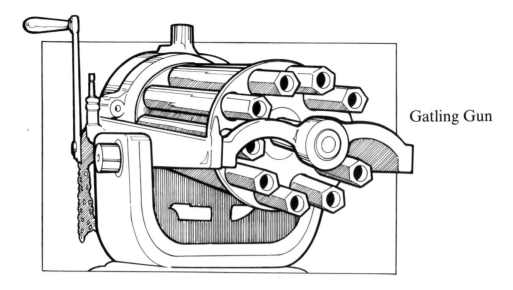

Gatling Gun

Hiram Maxim and the Machine Gun

THE MACHINE GUN (1884) SIR HIRAM MAXIM

The conduct of warfare was forever changed in the year 1884 by the invention of the Maxim Gun or, "Machine Gun," as it became commonly known. The weapon was able pour out a continuous and deadly stream of bullets onto the battlefield at a high rate of fire. It effectively ended the era of the mass cavalry or infantry charge.

An earlier version of an "automatic gun" was built during the American Civil War by engineer Richard Gatling. Its mechanism featured ten rotating barrels which were turned by a hand crank to fire ammunition from a gravity fed container, or "magazine." Although it had an impressive rate of fire (350 rounds per minute), it tended to jam often and was not put into widespread use.

The first truly automatic weapon was built by a prolific American inventor Hiram Maxim. By simply depressing and holding the trigger, his weapon was able to fire continuously from an ammunition belt which automatically fed bullets into the gun. In 1884 he successfully demonstrated his gun's capabilities to the British Military High Command. The British Army was suitably impressed and adopted the Maxim gun in 1889, with the Royal Navy following in 1892. The giant German industrial corporation, Krupps Armaments, obtained a patent license in 1893 and began to produce versions for other European countries.

By the beginning of World War I in 1914, all of the major combatants were equipped with machine guns. It was not until the awful reality of trench warfare, that the terrible firepower of this weapon was fully known. As a defensive weapon, it was able to cut down hundreds, even thousands, of troops if they tried to advance over open ground.

Although Hiram Maxim is best known for his development of the machine gun, he also had more benign inventions to his credit. Among these were the gas pressure regulator for gas lighting, the electrical pressure regulator for electric lights, a powerful headlamp for locomotives, an automatic sprinkler fire extinguisher, and even the curling iron.

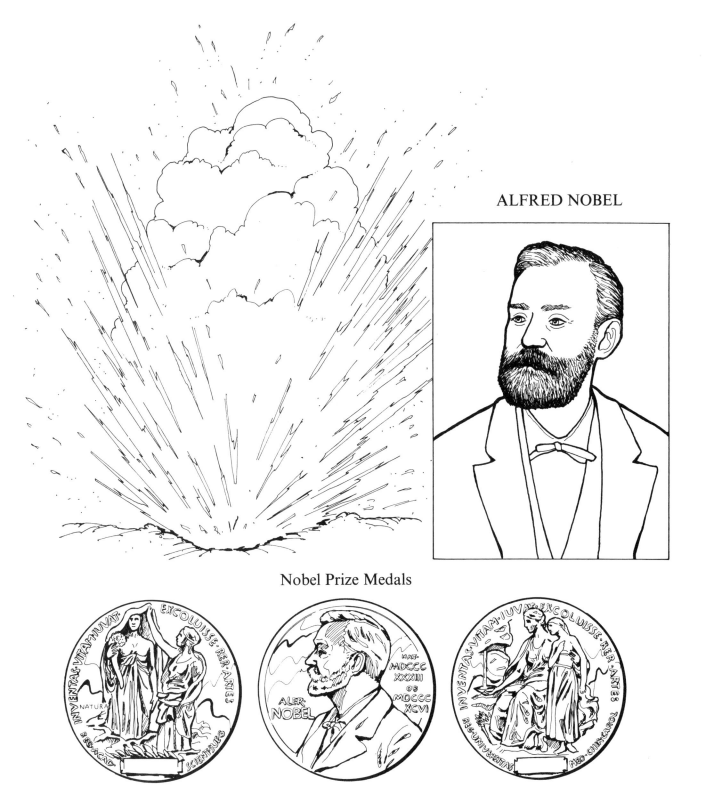

Nobel Prize Medals

DYNAMITE (1867) ALFRED NOBEL

Until the year 1846 the most powerful explosive known was gunpowder. In that year an Italian chemist named Ascanio Sobrero discovered a much more powerful compound, nitroglycerine. Because it was so unstable and sensitive, however, it was not practical to handle and use. In 1867, a Swedish inventor named Alfred Nobel (1833–1896), devised a method that rendered this powerful substance less dangerous to handle. He mixed nitroglycerine with a dough-like mineral called kieselguhr, and formed it into solid sticks. This enabled the explosive to be transported and detonated under controlled conditions. He called his explosive dynamite. He also invented another safety device for dynamite, a detonating "cap" from the chemical fulminate of mercury, now commonly called a "blasting cap." In 1875, Nobel invented an even more powerful explosive called gelignite. It was a mixture of nitroglycerine and nitrocellulose, also known as gun cotton. It was easier to use than dynamite, in addition to being more powerful.

Nobel's interest in humanitarian and scientific philanthropies led him to create a foundation for the purpose of bestowing awards on creative scientists and inventors. In 1901 he established the "Nobel Prize" for great achievement in the fields of Medicine, Chemistry, Physics, Economics, Literature, and ironically, though not surprisingly, for the promotion of International Peace. The Nobel Prize is considered the highest honor attainable in the fields in which it is awarded.

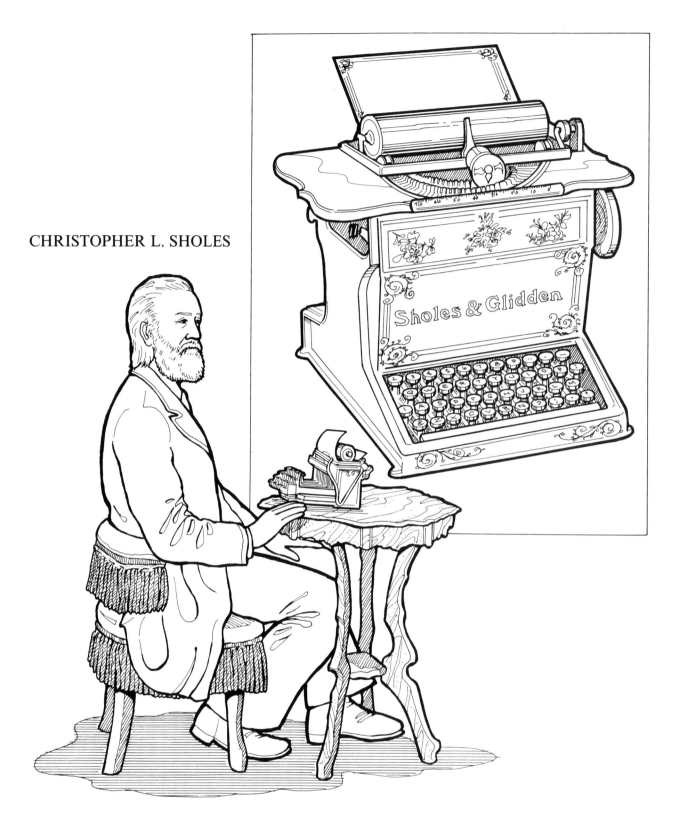

CHRISTOPHER L. SHOLES

THE TYPEWRITER (1868) CHRISTOPHER L. SHOLES

The "automatic writing machine," or typewriter, was invented by Christopher L. Sholes (1819–1890). He had already patented an "automatic numbering machine" in 1864. In conjunction with his partner, Carlos Glidden, they introduced this new writing machine in 1868.

In its basic principle of operation, his device used letter keys connected to type bars which struck an ink soaked ribbon, imprinting the letters onto paper. Because the keys and their type bars tended to jam, he invented a keyboard arrangement according to their frequency of use, rather than alphabetically. This system is still the standard arrangement on modern computer keyboards.

For the next five years Sholes modified his original machine with a number of improvements. In 1873 he sold his patent rights to the Remington Arms Company. The Remington name became synonymous with the commercial manufacture of typewriters. In 1883, American writer Mark Twain bought a Remington typewriter, and became the first writer to create an entire manuscript using this new writing tool. Although it was not an intended consequence, one of the most significant results of the typewriter was to bring large numbers of women into the work place as typists. The reasonably good wages paid to a typist allowed many women emancipation from strictly traditional roles.

ALEXANDER GRAHAM BELL

TELEPHONE (1876) ALEXANDER GRAHAM BELL

"Mr. Watson come here, I need you!" These were the words of inventor Alexander Graham Bell (1847–1922) to his assistant Thomas Watson, the first words ever transmitted by the telephone. Certainly one of the most important technical developments of the modern era, the telephone has become a universal convenience and necessity.

Its inventor was born in Edinburgh, Scotland, the son of a well-known teacher of the deaf. Young Alexander also became a teacher and advocate for the hearing impaired. In 1870 he moved to America and opened a school to train teachers to work with the deaf. Because of his interest and knowledge of acoustics, the science of sound, Bell began to experiment on a device to transmit sound over long distances using electricity. His invention was based on the principle that the sound vibrations of the human voice could be converted into specific electrical impulses by a transmitter, sent over an electrical wire, and reconverted back to the same sound vibrations by a receiver at the other end of the wire. The mechanism used to accomplish this was a skin membrane or diaphragm, much like a drumhead, with a vibrating piece of iron at its center. The voice of the speaker would cause the membrane and the iron to vibrate. These would be converted to electrical impulses by an electromagnet, and cause a similar device at the other end to vibrate with the pattern of the speaker's voice, creating audible and understandable speech. Alexander Graham Bell patented his telephone invention in 1876, creating a new era of communication. Bell's invention was widely accepted around the world, and by 1900 the telephone was a common business and household device.

Bell had a number of other inventions to his credit, among them the graphophone, a device to record sound, and a hydrofoil boat that captured the world water speed record in 1918. Along with his scientific research, Bell's work with the deaf continued to be an important factor in his life. With the royalty money from his inventions, he founded the Alexander Graham Bell Association for the Deaf.

THOMAS A. EDISON

Electric Light Bulb

PHONOGRAPH/GRAMOPHONE (1877), ELECTRIC LIGHT BULB (1879), ELECTRIC POWER STATION (1888), KINETOGRAPH (1891), and KINETOSCOPE (1894)   THOMAS ALVA EDISON

Perhaps the most famous inventor in history, Thomas Alva Edison (1847–1931) is a monumental figure in the development of American technology. For pure inventive genius and productivity, no one can compete with his astounding record of 1,093 patents. Most well-known for developing the electric light bulb in 1879, his inventions covered a wide range of technical fields.

Born in Milan, Ohio, he received only three months of formal schooling due to his partial deafness, but was taught at home by his mother, a woman of spirit, curiosity, and a love of learning. By the age of ten he had set up his own basement laboratory and had taught himself the basics of chemistry and electricity. His first working experience was as a telegrapher beginning around 1869. It was during this period that he built his first invention, the Edison Universal Stock Printer, for use by the New York Stock Exchange. The success of this device allowed him to open his first working laboratory in Menlo Park, New Jersey in 1876. Many of his most famous inventions were created in this laboratory, earning him the nickname "The Wizard of Menlo Park."

In 1877, his invention of the carbon-resistance transmitter greatly improved the newly developed Bell telephone. He also devised the first successful sound recording device, the phonograph or gramophone, in 1877. It consisted of a diaphragm and steel needle called a stylus, that recorded the sound onto a tinfoil covered cylinder. He then turned his attention to the creation of the incandescent electric light bulb. In 1879, after several years of research and experimentation with dozens of materials suitable for the glowing element, or filament for the bulb, he discovered that a carbon filament in a glass vacuum worked successfully. An English inventor, Sir Joseph Swan, also came up with a working light bulb in 1879.

In 1887 Edison opened a much larger laboratory in West Orange,

New Jersey. From this new factory he designed and built the world's first commercial electric power station. In 1888 Edison's Central Power Station, nicknamed the Pearl Street Station after its location in New York City, began to supply electricity for the users of his new light bulb.

In 1891 Edison invented the kinetograph, a motion picture camera that used George Eastman's newly developed flexible celluloid roll film. He also patented the first viewing machine for roll film called a kinetoscope in 1894. On April 14, 1894, Edison opened the world's first motion picture theatre at 1155 Broadway, New York City. It featured eleven kinetoscopes set up to view several movies including the first copyrighted movie, *The Sneeze*.

Thomas Edison's amazing number of inventions can be attributed to several factors. Primary among these of course was his intellect, Edison was driven by a powerful curiosity for how things worked but another critical aspect was his tremendous physical energy and stamina. He often worked twenty hours per day. In fact, Edison's most famous quote sums up his feelings about his abilities and achievements. "Genius," said Edison, "is 1% inspiration and 99% perspiration."

THE AUTOMOBILE (1895)   KARL BENZ

A German engineer, Karl Benz (1844–1929), is credited with being the inventor of the "horseless carriage," or motor car. In 1883 he established a factory to build small internal combustion engines in the town of Mannheim, Germany. His first motorized vehicle was the three wheeled buggy shown above. It featured a U-shaped frame with an engine that propelled the rear wheels with a chain drive. Benz successfully test drove his patent motor wagon around a cinder track near his factory in 1885. He sold the first of these vehicles in 1887 in Paris, and in 1893 he built his first four-wheeled automobile.

In 1885, the same year that Karl Benz test drove his motor car, another German engineer, Gottlieb Daimler, successfully built and drove the world's first motorcycle. It consisted of a wooden bicycle type frame with two large wheels and two small stabilizer wheels (what we would now recognize as training wheels). Daimler soon began building three and four wheeled motorized carriages in competition with Karl Benz. An intense rivalry between Daimler and Benz would last for the next 15 years as they contended for the honor of the inventor of the motor car. Eventually, Benz won out and was given credit as the originator. Ironically, their two companies merged in 1926 to form Mercedes-Benz, one of the world's largest and most prestigious auto makers.

Automobiles came into being in the United States in 1879 when a lawyer and inventor named George Seddon received a patent for his design of a motor car. Because he never actually built his vehicle, the credit for the first American made automobile is given to brothers Charles and James Duryea. They built their gasoline powered motor car in 1893.

The automobile has made such an impact on our economy and culture, that it is difficult to imagine a world without these machines. Dozens of industries were formed to support their development and widespread use, chief among these, the giant oil industry. The invention of the automobile has been the mechanism for an unprecedented era of mobility, convenience, and freedom, and their significance to world economy and culture has been immense.

ALTERNATING CURRENT (1885), and THE TESLA COIL (1891) NIKOLA TESLA

Nikola Tesla (1856–1943) was a brilliant and eccentric Croatian scientist. Tesla's first invention, built in 1882, was an electromagnetic induction coil that used the power of alternating current, or AC, rather than the more common system of direct current or DC, championed by Thomas Edison. Alternating current is so called because it constantly reverses its direction of flow and it is capable of generating much greater electric power.

In 1884 Tesla moved to the United States and began work at the Edison Laboratories as a research assistant. It was while at these labs that he developed an entire system for the transmission of electricity over long distances that used AC power. Edison vehemently opposed this AC system, fearing that it was too dangerous and uncontrollable. Tesla left the Edison Labs and sold his idea to American industrialist, George Westinghouse, who would use his resources to promote and implement Tesla's inventions. AC power eventually proved to be the superior system, and in 1895, a Tesla designed AC power station was chosen to harness the energy of Niagara Falls.

Tesla established his own laboratory where he invented the Tesla Coil. This device was a high frequency electrical induction coil that could be used to power long distance radio and television transmissions. By 1900 Tesla had theorized and envisioned a global communications network to transmit sound and pictures using his equipment. He was unable to get financial backing for this project, however, and it was never developed.

Tesla had a flamboyant ability and ambition to communicate his ideas, although he was very lax in pursuing patent rights for his concepts. In one of his more theatrical demonstrations, designed to show the safety of AC power, he read a book while sitting just a few feet from a high voltage coil generating "artificial lightning." He also showed how AC power was controllable by picking up wireless lamps and letting the current flow through his body to light the lamps. In another public presentation he demonstrated that radio waves could be used to power a light bulb.

In 1917 Tesla received the Edison Medal for his electrical research. He is remembered as a brilliant and enigmatic scientist whose many contributions were indeed ahead of their time.

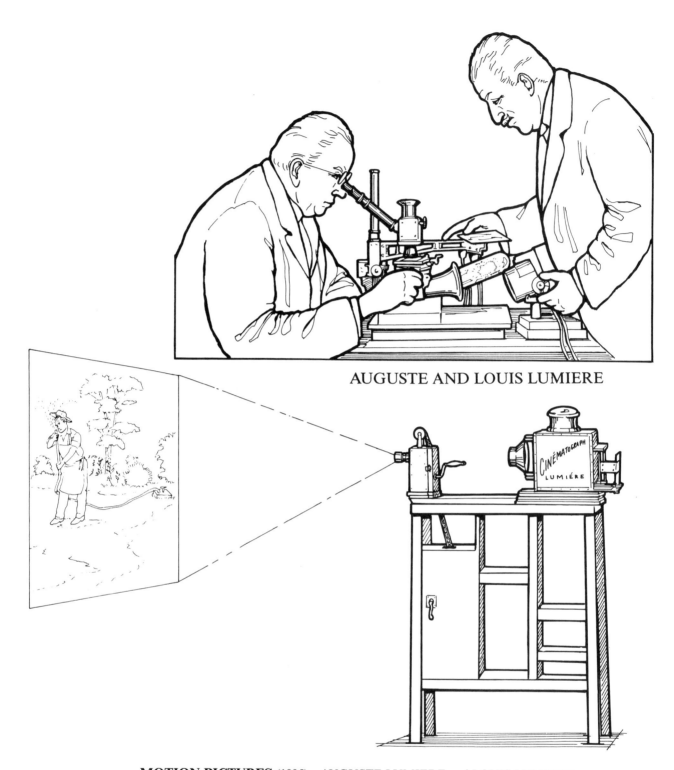

AUGUSTE AND LOUIS LUMIERE

MOTION PICTURES (1895) AUGUSTE LUMIERE and LOUIS LUMIERE

The premiere of the first projected motion picture took place on December 28, 1895, at the Grand Cafe in Paris. A series of ten short films lasting twenty minutes was shown. To the people in the audience, it did not seem particularly astounding. Little did they know how it would change the world. But just as painting was the dominant art form of the eighteenth century, and literature that of the nineteenth, motion pictures have become the most celebrated creative expression of our time.

Although many inventors and scientists were instrumental in the early development of moving pictures, including Thomas Edison, Eadweard Muybridge, J.A. Rudge, and J.E. Moray, it was two French brothers, Auguste and Louis Lumiere, who were the first to successfully demonstrate a machine to project movies. The Lumiere's owned an established and profitable photographic plate business and were keenly interested in the science of photography. This afforded them the opportunity to experiment with moving picture technology. After viewing one of Thomas Edison's kinetoscopes, the brothers built a machine that was both a motion picture camera and projector. They called this device the Cinematograph.

Despite the popularity and success of their moving picture machine, the Lumiere Brothers discontinued their work with movies after only a few more years of development. They felt that it was a curious novelty, but had no commercial potential. Other inventors and entrepreneurs saw otherwise, and by 1900, dozens of filmmakers were busily cranking out their own movies using the Lumiere system. Out of this early competition for the public entertainment market, came the movie studios of the twentieth century's filmmaking industry.

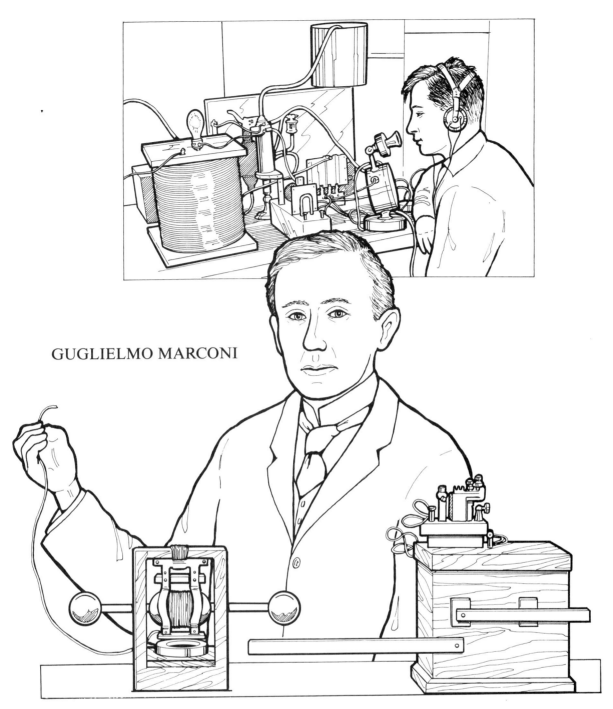

RADIO (1895)   GUGLIELMO MARCONI

Communications took a giant leap forward with the invention of radio, or wireless telegraph, as it was known at the time. Around 1865, two scientists working independently, James Clerk Maxwell in Great Britain, and Heinrich Hertz in Germany, had conceived the theory of the transmission of messages through the atmosphere by the generation of electromagnetic radiation or radio waves. A Russian scientist, Aleksandr Popov, was also instrumental in early radio research.

In 1895 a young Italian engineer named Guglielmo Marconi (1874–1937) took this theory and made it into a practical reality, by constructing a radio transmitter capable of sending a long distance signal. One of the innovative features of Marconi's equipment was his invention of the antenna. This device boosted the signal strength, enabling the longer distance transmission. It is still a vital part of radio technology. Unable to persuade the Italian government to support his further research, Marconi moved to England in 1896, where he was granted his first patent for wireless telegraphy. That same year he succeeded in sending a radio transmission across the English channel. This accomplishment enabled him to start his own company, The Wireless Telegraphy and Signal Company, in 1899.

At this time it was believed that the range of radio signals was not far, around 200 miles. It was thought that this limitation was caused by the curvature of the Earth's surface. In 1901 Marconi was able to send a radio signal from Cornwall, England, to St. John's Newfoundland, in Canada, a distance of over 2000 miles. Marconi had correctly theorized the existence of a layer in the Earth's atmosphere that would bounce back radio waves to increase their range. We now know this atmospheric layer as the ionosphere.

Marconi continued his work on radio technology, building improved and more powerful transmitters, throughout the first two decades of the twentieth century. For his achievements in radiotelegraphy, Guglielmo Marconi was awarded the Nobel Prize for Physics in 1909. His company, Marconi Wireless Telegraph and Signal, is still in existence as a leading manufacturer of electronic equipment.

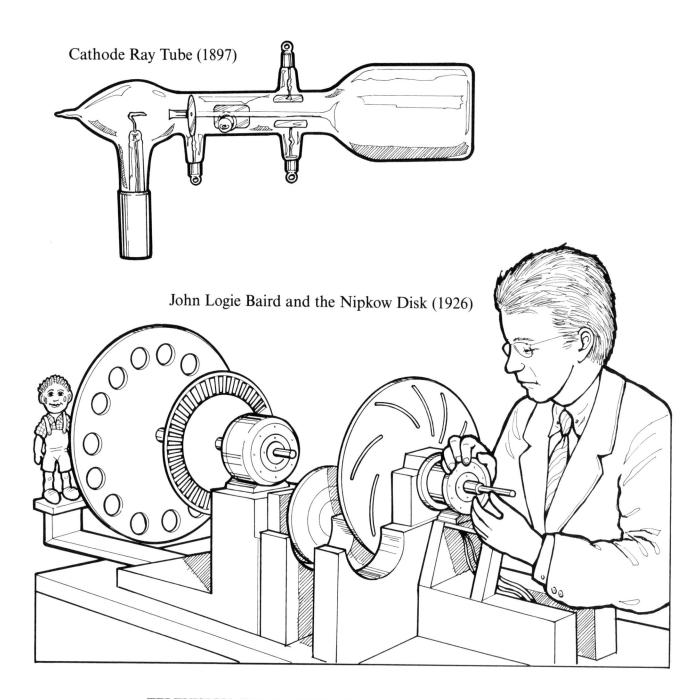

Cathode Ray Tube (1897)

John Logie Baird and the Nipkow Disk (1926)

TELEVISION: THE CATHODE RAY TUBE (1897)   KARL BRAUN
FIRST TV TRANSMISSION (1925)   JOHN LOGIE BAIRD
ICONOSCOPE, KINESCOPE (1923/24)   VLADIMIR ZWORYKIN

Historians and scientists might debate the relative merits of a number of inventions, but in terms of having the most profound impact on the greatest number of people, television would have to be considered one of the most influential. This amazing device has turned a planet with hundreds of nations and races into a Global Village by its incredible power to inform, entertain, and educate.

A number of scientists and inventors had a hand in the creation of television. Its roots can be traced to a device invented by an English scientist, William Crookes in 1887. He constructed a glass tube within which he placed two metal electrodes. When the air was removed from the tube and an electric current passed through the electrodes, the tube glowed. Crookes called these glowing electrons cathode rays. In 1897 a German scientist named Karl Braun (1850–1918) went a step further. He placed a metal plate at one end of a cathode ray tube. The plate was covered with a chemical that glowed when struck by the cathode rays or electrons. By varying the intensity of the electric current, Braun was able to create a moving spot of light on the coated plate. This is the basic principle of the modern television picture tube.

Two distinctly different approaches were taken in the early years of television development. One was electromechanical, pursued by Scottish inventor John Logie Baird (1888–1946), and based on a system utilizing "Nipkow Disks." The other approach was completely electronic, developed by a Russian immigrant to America, Vladimir Zworykin (1889–1982), and was based on the cathode ray tube.

Baird's electromechanical system relied on a series of spinning disks that optically transformed the image of an object into electrical impulses. These disks were invented by German engineer Paul Nipkow in 1884. Using this technology, Baird transmitted the first

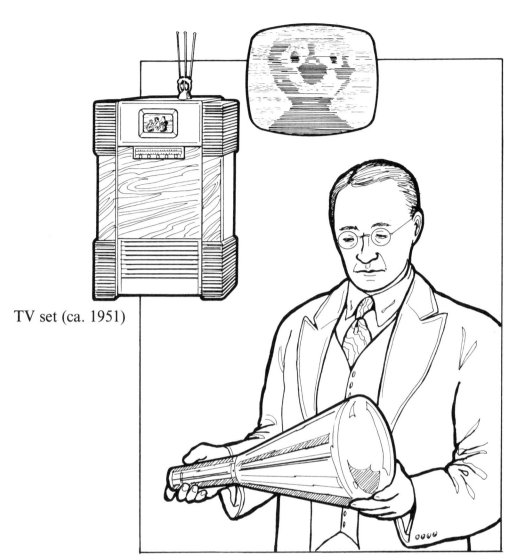

TV set (ca. 1951)

Vladimir Zworykin and the Electron Camera Gun

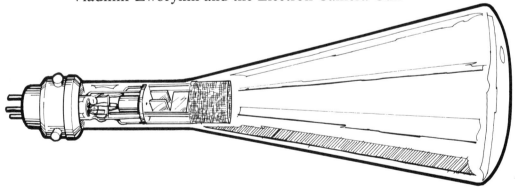

television image on October 2, 1925. In 1927, he was able to send a long distance transmission from London to Glasgow, Scotland. An even farther transmission was accomplished in 1928, with an image sent from London to New York. The quality of the images using the Nipkow disks was poor, however, and it was not developed much past 1935.

At the same time as Baird was experimenting with his electro-mechanical television, Vladimir Zworykin was working on his electronic system. In 1923 he perfected and patented the electron camera scanner, which he called the iconoscope. He also received a patent in 1924 for a television receiver, which he labeled a kinescope. Zworykin's iconoscope used a beam of electrons directed at a coated screen that was broken up into thousands of tiny squares. These individual squares glowed when hit by the electron beam. The images created on the coated screen were transformed into electrical impulses. These were then transmitted in the same manner as radio signals. At the other end, a receiver, the kinescope reversed the process, turning the electrical impulses into tiny spots of light. Since the human eye cannot see or follow these spots of light individually, it translates them into larger patterns that make up the recognizable images we see on the TV screen. The clarity of the image using this electronic system was far superior to Baird's approach. Zworykin's work, along with that of another American television pioneer named Philo T. Farnsworth, laid the foundation for modern television technology.

By 1936, the first public television broadcast was made in England. In the United States, the RCA company set up the first regular television broadcasting network in 1939.

WILBUR and ORVILLE WRIGHT

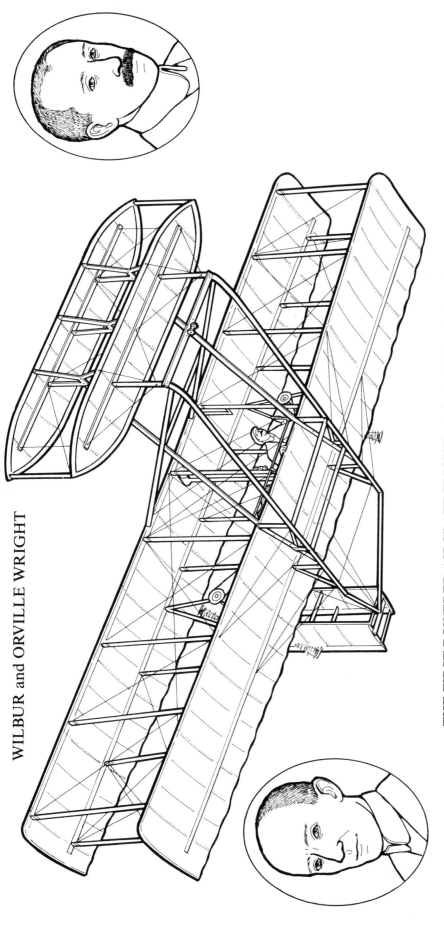

THE FIRST POWERED AIRCRAFT FLIGHT (1903)   THE WRIGHT BROTHERS

The age of powered flight began at the windswept Kill Devil Hills, near Kitty Hawk, North Carolina, in the year 1903. On December 17th, American inventor Orville Wright (1871–1948) piloted the first propeller driven aircraft for a flight lasting only twelve seconds.

The era of modern flight begins in the 1880s, when the concept of the airfoil wing, the main lifting mechanism for aircraft, was fully understood and developed by German aviation pioneer Otto Lilienthal. He successfully built and flew a series of unpowered gliders based on the airfoil principle.

The Wright Brothers, Wilbur and Orville, were owners of a successful bicycle repair business, expert mechanics, and had a passion for flight. Starting in 1899, they built a series of kite like biplane gliders (one wing on top, one on the bottom) from wood and fabric. By 1902 they had designed an aircraft that was very stable and controllable. To power this machine, the Wright Brothers fabricated their own lightweight gasoline engine. It developed twelve horsepower from its four cylinders, and drove two hand built pusher propellers by means of a bicycle chain. A pusher propeller is mounted at the trailing edge of the wing, and pushes the plane forward through the air. This is in contrast to the more common type of propeller system, the puller, which is mounted at the leading edge of the wing, and pulls the aircraft ahead. Their aircraft was christened the *Flyer I*.

The Brothers made an initial attempt at a powered flight on December 14th, but were unsuccessful. Three days later, on December 17th, Orville Wright piloted their first historic flight. They made three additional test flights that day. The longest lasted fifty-nine seconds and was flown by Wilbur Wright (1867–1912). In 1904, they built the *Flyer II* and demonstrated its controllability with aerial maneuvers that included banked turns and landings that returned to the take off point.

Amazing as it may seem, the Wright Brothers' accomplishments received little public attention in the United States. It was not until Wilbur traveled to Europe to demonstrate their *Flyer* that they began to receive widespread acclaim. The Brothers continued to develop improved flyers, and in 1909, the U.S. Army ordered the first military versions of their aircraft. Within a few short years many inventors and manufacturers, in both the U.S. and Europe, were producing their own aircraft designs. From the simple wood and fabric Wright Flyer, with its flight of just twelve seconds, came one of the wonders of the modern world.

THE VACUUM CLEANER (1901)   HUBERT BOOTH and MURRAY SPENGLER

Not all inventions are great and profound. Some are just handy and convenient. They become part of our daily lives by virtue of their utility. Such is the vacuum cleaner. It was initially devised in 1901 by an English mechanic named Hubert Booth. He constructed an electrically powered machine that sucked up and filtered dust and debris. The device itself was so large and cumbersome that it had to be pulled from building to building by a team of horses. Needless to say it never became a popular appliance.

In the United States meanwhile, another inventor named Murray Spengler built a much smaller and more portable electric vacuum system. He was awarded a patent for his device in 1908 but lacked the funds for further development. He sold his patent rights to prominent businessman William Hoover who began the manufacture and sales of the machines.

They soon became quite popular for both home and industrial use. The Hoover Upright Model "O" vacuum cleaner rolled on wheels, had an electric fan to create the suction, and a dust collector bag mounted on a long handle. Even though these early model household vacuum cleaners were much smaller than Booth's original horse-drawn version, by today's standards they were still bulky. Two people were often required for their efficient use. The vacuum cleaner was one of the first of many household appliances that would make life in the twentieth century more convenient and comfortable.

LEE DE FOREST

THE TRIODE RADIO VACUUM TUBE (1906)   LEE DE FOREST

In the first decade of the twentieth century, radio communication was limited to telegraphic-type coded signals. An American inventor and radio engineer, Lee De Forest, played a crucial role in the development of voice transmissions by radio waves. He improved on a earlier device called a diode radio vacuum tube, so called because of the two electrodes within the tube. The diode was invented by Sir Ambrose Fleming in 1904, who called his device the thermionic radio valve. De Forest's modified version was called the audion tube. Both of these inventions allowed a weak form of voice transmission by radio.

In 1906 De Forest made a significant breakthrough in radio technology. He added a third electrode to the diode, making it a triode. This tube strongly amplified radio voice signals by locking on to a "carrier wave." It proved to be a vital element in the further development and widespread use of radio for regular voice communications.

Lee De Forest (1873–1961) continued his pioneering work in radio and patented over 300 inventions during his career. He proved that sometimes a small device can have a very large impact. His triode tube was an indispensable component in radio technology until the introduction of its successor, the transistor in 1947.

MASS PRODUCTION and THE AUTOMOBILE (1908) HENRY FORD

America at the turn of the century was a nation heavily dependent on horses for most of its basic transportation needs, but by developing the assembly line, mechanic turned industrialist Henry Ford (1863–1947) was able to produce automobiles in a quantity, and at a price, that made them available to the average working person.

Henry Ford was the son of Irish immigrants, and like many children of that era, received little formal education. He showed a keen interest and aptitude for mechanics, however, and by 1899 had become Chief Engineer at the Detroit Edison Company. His interest in mechanics prompted him to build his first automobile in 1892. He began his own automobile company in 1903.

The mass production system used by Ford for his automobiles was pioneered by other American inventors including Eli Whitney and Samuel Colt. The basic principle of the system was for the manufacturing process to be broken down into smaller tasks along an assembly line. As the product traveled down the line it would have parts or systems progressively added until reaching completion at the end of the line.

In 1908 Ford began mass production of the famous "Model T" or "Tin Lizzie." By 1913, most working Americans could afford the $500.00 price of a Model T. With as many as 1000 cars per day being built, Ford sold over 15 million Model T's between 1908 and 1927. By the end of their production run, the cost of a Model T had been lowered to just under three hundred dollars.

Besides being an innovative businessman, Henry Ford was an enlightened employer who steadily increased his workers' pay, reduced their hours, and instituted a profit-sharing plan. His success in business made him a very wealthy man. This in turn enabled him to establish a charitable organization, the Ford Foundation, to donate money to worthwhile individuals, causes, educational institutions, and others. The automobile industry, as represented by Ford, did much to pave the way for the economic growth and security of the U.S. during the first few decades of the twentieth century.

E.D. SWINTON

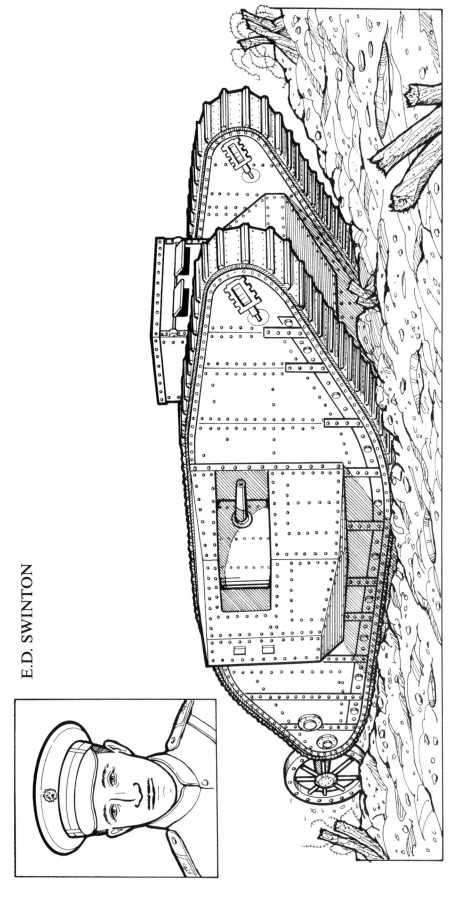

ARMORED, TRACK-DRIVEN BATTLE TANK (1916)
LIEUTENANT COLONEL E.D. SWINTON, LIEUTENANT W.G. WILSON, and W. TRITTON

The armored, track-driven battle tank, was devised by the British and first used in combat in 1916. Interest in development of the tank, or landship as it was called then, was prompted by Winston Churchill in 1915. At that time he was the first Lord of the Admiralty, our equivalent of Chief of Naval Operations. He became aware of the investigation by Lieutenant Colonel E.D. Swinton (1868–1951) of the possibility of creating an armored fighting vehicle, a landship, using an American Holt tractor propelled by continuous "caterpillar tracks."

In November 1915, Churchill ordered Royal Navy Lieutenant W.G. Wilson (1874–1957) and engineer W. Tritton (1875–1946) to construct such a vehicle using a Holt tractor and boilerplate for armor. Although this prototype was not very successful, it led to the further development of a new tank, the *Mark I*, nicknamed "Big Willy." This vehicle was the first to utilize the classic tank shape. It was twenty-six feet long, weighed twenty-eight tons, and could travel at four mph. It was armed with either a cannon, or several machine guns. The most important features being: it was impervious to machine gun fire, could cross a 9 foot-wide trench, and could breach treacherous barbed wire fortifications.

By September, 1916, the British had sixty tanks ready for combat during the Battle of the Somme. It was not an auspicious beginning. Of the forty-nine tanks assembled at the staging area, thirty-two started the battle, and only nine crossed the no-man's-land of the battlefield. In a subsequent engagement, the Battle of Cambrai in 1917, tanks proved their worth. Four hundred British *Mark I* tanks turned the tide of battle and led to the capture of 30,000 German prisoners and 800 field cannons.

The Germans built only a few tanks during World War I to contest the British tank efforts. It was during the inter-war years of the 1930s that the German Army made great strides in tank development, with a series of advanced battle tanks. These proved to be powerful and formidable weapons that played a key part in the Blitzkrieg, or Lightning War, conducted by the German Army during the early years of World War II.

RICHARD H. GODDARD

THE LIQUID-FUEL ROCKET (1926)   ROBERT H. GODDARD

The foundation for modern rocket science and space flight was begun on a Massachusetts farm, March 16, 1926. On that day, American physics professor Robert H. Goddard (1882–1945) successfully launched the world's first liquid-fuel rocket. It flew for only two and one-half seconds to an altitude of just 184 feet, but it demonstrated the principle that rocket flight was possible and practical.

The rocket propulsion system itself was several hundred years old. The British Army used rockets in warfare as early as 1845. These rockets were propelled by gunpowder charges but were unreliable, inefficient, and could carry only a small payload or warhead. Goddard felt that a better method of rocket propulsion was possible and began his study and experimentation while teaching at Clark University, in Massachusetts. He theorized that the ignition of two combustible liquids would provide much greater power and thrust than a solid propellant, such as gunpowder. His design for a rocket engine mixed liquid oxygen and gasoline in a reinforced combustion chamber. When they were ignited, the explosion and resultant expansion of gases were channeled out a bell shaped opening at the rear of the chamber. This created a powerful forward thrust to launch the rocket.

Goddard continued working on progressively larger and more powerful rockets throughout the 1930s and 40s. By 1941 he had launched a four stage rocket (shown above) that reached an altitude of 10,000 feet and a speed of over 700 mph. In his experimental rockets, he was the first to incorporate many of the features of modern spacecraft including gyroscopic stabilizers, and steering fins to guide the rocket.

During his lifetime Goddard received no government support for his pioneering work. With the passage of time, we are able to recognize the direct impact his work had on the development of spacecraft like the *Saturn 5* moonrocket and the Space Shuttle. Goddard died in 1945 without seeing how important his research was to the future of space flight. As a belated tribute, the National Aeronautics and Space Administration (NASA) named its Goddard Space Flight Center in Maryland after him.

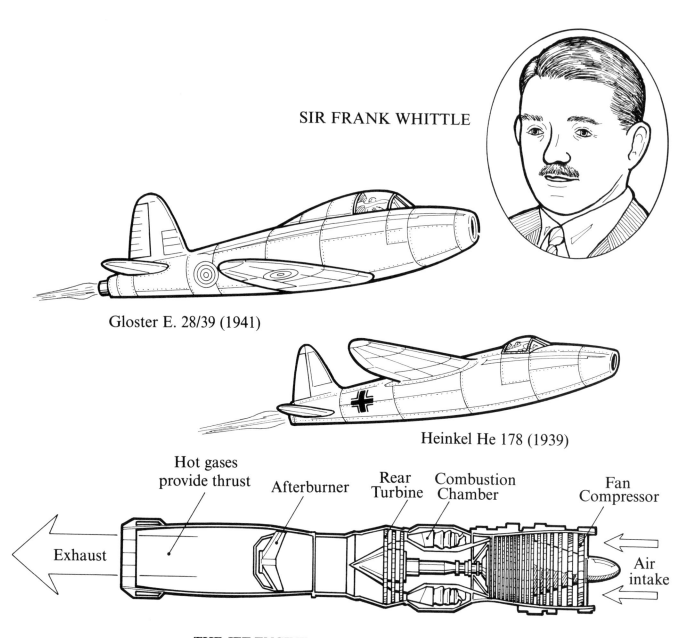

SIR FRANK WHITTLE

Gloster E. 28/39 (1941)

Heinkel He 178 (1939)

THE JET ENGINE (1937)

The first patent for a working jet engine was granted in 1930 to a young Royal Air Force Lieutenant, Frank Whittle (1907–1996). Whittle conducted his research and experimentation while assigned to the Royal Air Force College. His jet engine design consisted of a fan like compressor which sucked air into the front of the engine while also compressing and heating it. The air was then drawn into a combustion chamber where it was mixed with gasoline and ignited. The resulting burning gases were exhausted rearward providing the propulsive thrust. As the gases were ejected they also turned a turbine wheel which was connected to the front compressor. This kept the engine cycle in operation. Whittle continued development of his engine throughout the 1930s, and by 1937 had successfully tested a working prototype. At that point, due to minimal British government interest and support, it was still an engine without an aircraft. That was not the case in Germany, however. The German Luftwaffe (Air Force) was gearing up for World War II and showed great interest in this new type of aircraft engine. In 1936, German engineer Hans von Ohain began work on a jet engine based on Whittle's design. After a series of successful test flights, several German aircraft companies began development of an operational jet-powered fighter aircraft. The most prominent of these was the Messerschmitt *Me 262 Swallow* twin-engine fighter, first flown in combat in 1944.

SIR FRANK WHITTLE

The British Royal Air Force also tested and operated a jet fighter during World War II. By 1941 they had mated one of Frank Whittle's jet engines with a prototype aircraft designated the *E. 28/39*, and had begun conducting test flights. By 1944 they had developed this prototype into the operational twin-engine Gloster *Meteor Mark I* fighter plane.

The first U.S. Air Force jet fighter to fly was the Bell *P-59 Airacomet*, test flown in 1942. It was used mainly to gather flight data for this new propulsion system and never put into widespread operational use. Its successor, however, the Lockheed *P-80 Shooting Star*, was deployed in Italy during the last few months of the war, but did not engage in combat. It would later prove to be a valuable aircraft in the Korean War starting in 1950.

Jet engine technology has advanced rapidly since the 1940s. These powerful and reliable engines now propel almost all commercial and military aircraft, including helicopters. For the last twenty-five years, the world's fastest and highest flying jet aircraft has been the twin-engined American reconnaisance plane, the Lockheed *SR71 Blackbird*. It has a top speed of over 2,200 mph and can reach an altitude of 80,000 feet.

CHESTER CARLSON

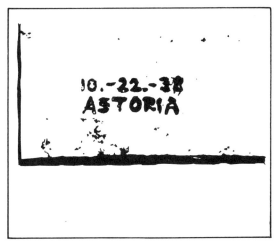

First xerographic image

Photocopier Prototype (1938)

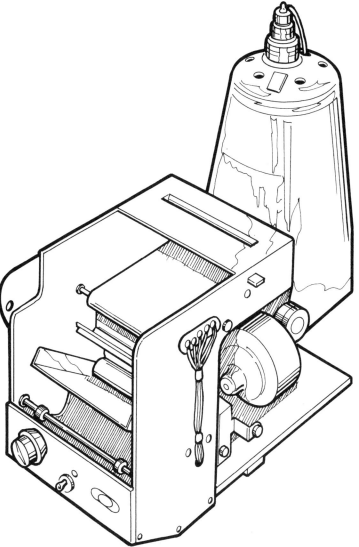

XEROGRAPHY (1938) CHESTER F. CARLSON

The xerographic copier was devised by American physicist Chester F. Carlson (1906–1968), and uses a process based on the principle of electrophotography. The name xerography is derived from the Greek words *xeros,* meaning dry, and *graphos,* meaning writing. That is a reference to the dry ink or toner used in the process to create an image on paper.

Carlson began his experimentation while employed in the patent department of a large New York electronics firm. Working alone and in his own home, he discovered that the element selenium could be used as a transfer conductor for creating an image with electrically charged black powder, or dry ink. Once the image was transferred from a selenium drum, the powder image was hardened or fused onto paper by heat and chemical vapors. On October 22, 1938, Carlson successfully printed the first xerographic image.

He was awarded a patent for his process in 1940 and began a search for the financial and technical support he needed to build and sell his copier device. In 1947, Carlson licensed the manufacturing and marketing rights for his invention to the Haloid Company, of Rochester, New York. This small photographic paper company eventually became the copier giant, Xerox Corporation, with the successful introduction of their model 914 fully automatic copier in 1959.

Today, many companies manufacture copiers ranging from simple desktop models to room size, high speed duplicators able to churn out hundreds of copies per minute. Chester Carlson's xerographic copier has become a valuable and integral part of the modern business world.

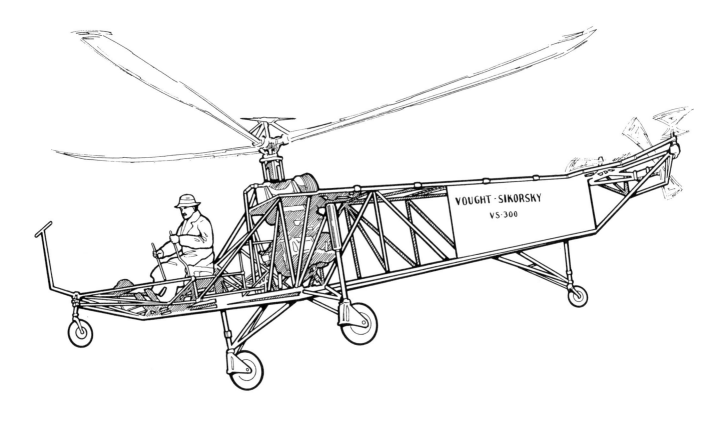

LEONARDO DA VINCI

IGOR SIKORSKY

THE HELICOPTER (1940)   IGOR SIKORSKY

Several pioneers of aviation had a hand in the development of the helicopter, but one, Igor Sikorsky (1889–1972), stands out as the driving force. Sikorsky was a Russian-born American immigrant, and one of the key figures in modern aviation.

Sikorsky emigrated to the United States after the Russian Revolution of 1917. He founded Sikorsky Aero Company which later became a subsidiary of the much larger United Aircraft Company.

Sikorsky was always interested in developing an aircraft capable of vertical take off and landing. He was influenced by the work of Spanish aeronautical engineer Juan de la Cierva, inventor of the autogyro in 1937, and by the helicopter prototypes built by the Focke Achgelis Company in Germany. The autogyro was a curious cross between a conventional airplane and a modern helicopter. It had a nose mounted propeller and upward lifting rotor blades but could not hover in mid-air–a characteristic of true helicopters.

Sikorsky's main difficulty in building a helicopter was overcoming the problem of torque. This is the property whereby the body of the helicopter turns in the direction opposite to the direction of rotation of its rotor blades. Sikorsky solved this by mounting two smaller, vertical rotor blades at the tail of his helicopter which counterbalanced the torque. In May 1940, with Sikorsky himself as the test pilot, his *Model VS-300* helicopter took off vertically, successfully flew for fifteen minutes (including hovering and landing maneuvers). A new age of flight had begun. The U.S. Army was impressed enough with the *VS-300* that it ordered 400 aircraft to be delivered in a military version called the *XR-4*. Beginning in 1943, they went into limited service during the Second World War.

It was during the Korean War in the early 1950s that the helicopter really proved its worth as a military observation aircraft and rescue vehicle. The Bell *Model 47*, the familiar bubble nosed "whirly-bird," proved to be a life-saving instrument in the quick transport of wounded soldiers from battlefields to medical facilities. In the Vietnam War of the 1960s and 70s, the helicopter was armed and became an important military weapon.

JACQUES YVES COUSTEAU

SCUBA UNDERWATER DIVING SYSTEM (1943)   JACQUES-YVES COUSTEAU, EMILE GAGNAN

Seven-tenths of the Earth's surface is covered by water. Within the liquid depths of our great oceans, seas, and lakes lies a world teeming with life. The final step in discovering this underwater world was the development of SCUBA diving gear. SCUBA is an acronym of the words, Self Contained Underwater Breathing Apparatus. It is also commonly called the Aqua-Lung.

It was invented by oceanographer and underwater explorer, Captain Jacques-Yves Cousteau (1910–1997). He began his career as a French Naval Officer working on underwater diving technology during World War II. In 1942 he began research with French engineer Emile Gagnan on a system that would allow a diver to carry an air supply along with them in their underwater work. This would free them from the restriction of a clumsy air hose to the surface.

The key to this system was the air-demand valve, now called the regulator. This device allowed the diver to breathe compressed air from a tank on her back, and to exhaust the carbon dioxide expelled from her lungs. When in June 1943, Cousteau and Gagnan successfully tested their SCUBA equipment to a depth of sixty feet, an important new tool was added to the quest for knowledge about the oceans.

Captain Cousteau would later go on to become a world renowned oceanographer and underwater explorer. His dozens of award winning books, films, and television shows have contributed enormously to our understanding of the immense underwater world, and the Aqua-Lung has allowed divers to engage in their own research, exploration, and recreation.

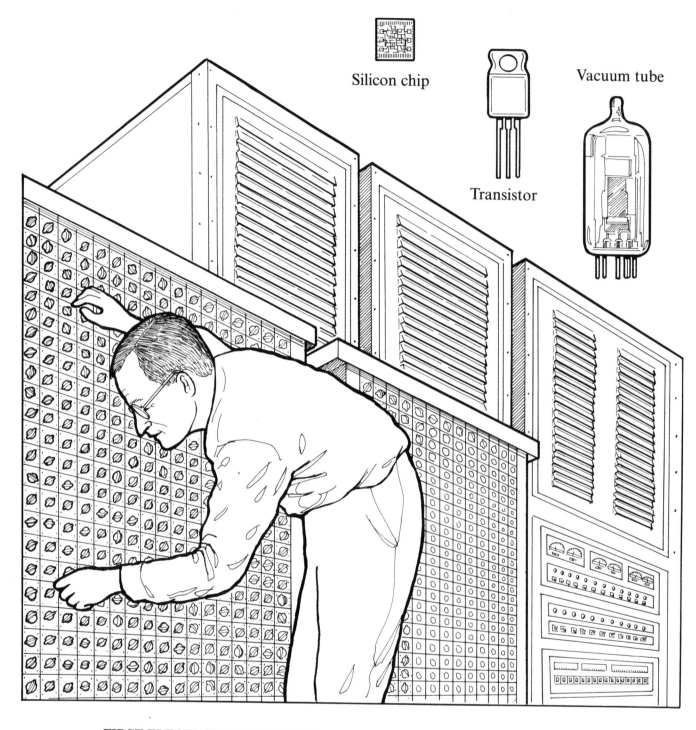

FIRST ELECTRONIC COMPUTER (1945) JOHN ECKERT and JOHN MAUCHLEY

As inventions have become more and more complex, they have become increasingly the product of teams of scientists and engineers, rather than the result of an individual effort. These groups of technical professionals usually work within corporate or academic research institutions. Such is the case in the development of modern computers.

The forerunner of today's powerful personal and office computers was built in 1945. It was christened "ENIAC," an acronym for "Electronic Numerical Integrator and Computer." It was built by engineers John Prospect and John Mauchley working at the Moore School of Electrical Engineering of the University of Pennsylvania in Philadelphia. By today's standards it was a huge machine. It filled a large room, 80 by 30 feet, and weighed an estimated thirty tons. It contained over 18,000 vacuum tubes and 170,000 resistors.

This equipment generated a great deal of heat, a problem that would plague ENIAC and other early computers until the invention of the transistor made the vacuum tube obsolete.

ENIAC's computing power was a miniscule 5000 calculations per second. That is in comparison to the power of our current crop of computers that have the ability to perform 10,000 megaflops, or 10,000 million operations per second. The invention of the transistor and the silicon chip has enabled this staggering increase in power. The scientific team of John Eckert and John Mauchley was also responsible for another well known early computer, UNIVAC (Universal Automatic Computer), built in 1950. During the presidential election of 1952, the UNIVAC computer correctly predicted the victory of President Eisenhower with just 7% of the votes counted.

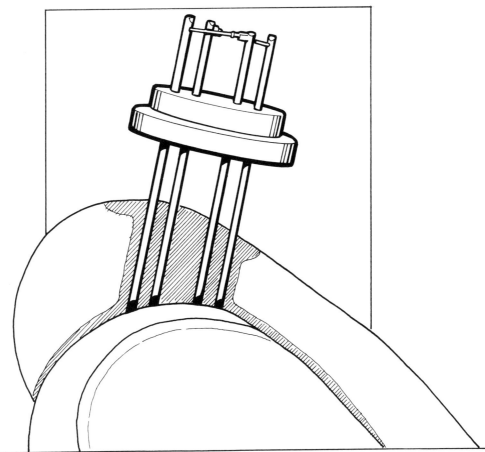

THE TRANSISTOR (1947) WILLIAM SCHOCKLEY, JOHN BARDEEN, and WALTER BRATTAIN

A crucial device enabling the advanced electronics of the information age is the transistor. This invention replaced the bulk of the vacuum tube—a critical part in all electronic equipment. It consists of a solid crystal of an element such as silicon or germanium, enclosed in a metal casing about the size of a sugar cube. Materials such as silicon and germanium are neither conductors nor resistors of electric current, but of a different class called "semi-conductors." They work by controlling the flow of electrons through the solid structure of the crystal, rather than through the airless vacuum of a tube.

Transistors have many advantages over old style vacuum tubes. They are smaller, more reliable and rugged, require less power to achieve the same result, and don't generate the damaging heat of a vacuum tube. This remarkable innovation was developed in 1947 by the Bell Laboratories scientific team of William Schockley, John Bardeen, and Walter Brattain. It has played a key role in contributing to the convenience and reliability of miniaturized electronic devices like radios, televisions, computers, telephones, and industrial equipment.

CHARLES HARD TOWNES

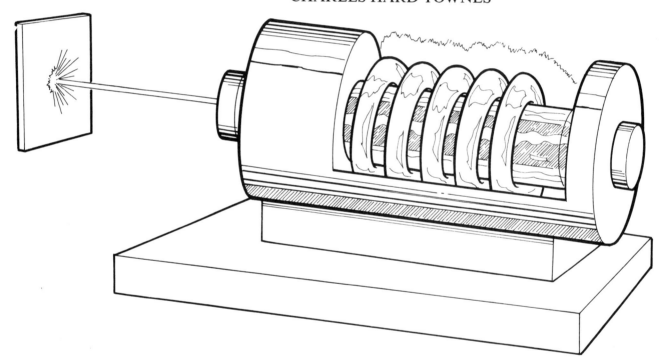

THE LASER (1960)   CHARLES HARD TOWNES, THEODORE MAIMAN

The laser is one of those inventions that gradually grew into a device with an incredibly wide variety of applications. Its basic principle is the use of a powerful and concentrated beam of light where all the light waves are "in step" or traveling in the same sequence. The term laser is an acronym for Light Amplification through Stimulated Emission of Radiation.

It was developed from an earlier communications innovation called the maser, built by American physicist Charles Hard Townes in 1954. The word maser stands for Microwave Amplification by Stimulated Emission of Radiation. Townes discovered that microwaves could be generated in an intense and undeviating beam that could provide long distance communication when directed at a specific receiver. His mechanism for generating this highly focused beam was to use the "cascade effect" to generate intense, overlapping non-scattering microwaves.

The maser proved to be a useful communications tool, and in 1957, Townes began work on a similar system using light waves instead of microwaves, calling it the laser. In 1960, Theodore Maiman, another scientist using Townes research as a guide, built the first laser. Maiman's device used a synthetic ruby rod with a helical flashtube wound around it. When the flashtube fired, the molecules in the ruby rod were stimulated to emit light. This reflected back and forth through the crystal from mirrors located at both ends of the rod. These were then amplified into a concentrated, narrow beam of light having parallel light waves.

These intense beams of light have become the operating mechanism for hundreds of products and devices. Lasers are used in communications, electronics, microsurgery, industrial processes, computers, and a myriad of other applications.